MARKET GARDENING

MARKET GARDENING

RIC STAINES

The Crowood Press

First published in 1990 by
The Crowood Press
Ramsbury, Marlborough,
Wiltshire SN8 2HE

British Library Cataloguing in Publication Data

Staines, Ric
  Market gardening.
  1. Horticulture
  I. Title
  635
  ISBN 1-85223-241-2

This book is dedicated to all who make a living from the soil

Line-drawings by Claire Upsdale-Jones

Typeset by Alacrity Phototypesetters, Banwell Castle,
Weston-super-Mare
Printed in Great Britain by Butler & Tanner Ltd, Frome

Contents

Acknowledgements

I would like to thank all those who have helped me in writing this book, especially my colleagues at Otley College, the students, my parents who gave me an interest in horticulture, and last but not least my long-suffering wife and children without whose help, encouragement and typing it would not have been possible. I would also like to thank Otley College and the following people for allowing me to photograph their holdings, farm shops, etc: J. Coles of Roseleigh Nurseries, Kirton, J. Craig of Nature's Foods, Little Bardford, R. Blyth of Friday Street Farm Shop, Farnham, Saxmundham and A. Simpson of Hillside Nurseries, Hintlesham.

Introduction

Market gardening. What do I mean by market gardening? When I put this question to a colleague he said, 'There's no such thing now.'

'Yes there is,' I replied, 'it's ... it's ... well, you know, a market garden.' These two words hold a lot of meaning, much more than just 'gardening for a market'. For this is really how it all started, with people selling off the surplus produce from their gardens for a little extra. As towns such as London grew bigger, a steady market for fresh food developed.

Market gardening in England properly started in the sixteenth century. The charter for the Gardeners' Company and the Fruiterers' Company was granted in 1605. It was said that a holding of 3 acres (1.2 ha) could keep a man, his family and employ outside labour. In 1750 bell-glass cloches and garden frames were being used and produce was being grown and sold close to roads direct to the public. Was this the birth of the farm shop? The market garden grew from these beginnings. Always just outside the town and moving out as the town moved out, they provided a range of vegetables, fruit and flowers to town and, later, to city dwellers. The range increased as new types were introduced from abroad and gained acceptance with the buying public. This development has continued right up to the present day. We tend to think market gardening has died out because new terms have appeared: smallholding, self-sufficiency, etc. These are really trying to describe something different. So what is a market garden? By tradition, it is a smallholding of up to 10 acres (4 ha) or so producing a range of crops such as vegetables, fruit and possibly a few flowers. Modern influences have changed this, as I hope this book will show.

The book will also give the reader a good understanding of the principles of growing crops commercially in this day and age. It is not a blueprint for how to grow certain crops, but covers a comprehensive range of factors to ensure that anyone in, or thinking of taking up market gardening will stand a better chance of success. Good luck to you.

1 Market Gardening Today

The modern market garden holding is very different to its predecessors. This is due in no small part to the many modern developments within the horticultural industry. Improved varieties of seed, methods of propagation and growing, and the advances made in plant nutrition have all had their impact. So too has the change in the traditional outlets for the produce from the market garden. Both the buying and eating habits of the public have changed dramatically since the war. In this ever-changing world the market garden has had advantages. It is perhaps easier for small producers to alter or adapt their ways. More recently, the changes in the way consumers buy their fresh produce have made life both easier and more difficult. The advent of supermarkets has made it virtually impossible for small growers to supply them. This, coupled with the reduction in the number and importance of local wholesale markets, has meant changes to the outlets for market gardens. Some have found success in specialising in certain crops, in particular those crops which require a high degree of skill and labour, and so are not as profitable for the big extensive growers to produce. The profit margins on crops such as lettuce have not risen in real terms for at least the past ten years so the grower has either to become more and more efficient and increase levels of production or find a more profitable line. All these factors combined have helped to make the market gardener specialise.

More recently, the interest in wholefoods and organically produced food and the advent of the farm shop and pick your own have revitalised the market garden. Farm shops grew out of a need for the small producer to maximise his returns by cutting out the middlemen. Development has now virtually turned full circle as many farm shops have really become out of town shops, not large and supplying everything, but specialising in fresh local produce. Some also supply imported produce such as oranges and bananas and even more exotic fruit because this became a natural extension of the business. Pick your own developed alongside this as no harvesting labour was required - the public did it themselves.

More recently again, the swing of public attitudes to care for the environment and to move away from a prepackaged life-style should go far in aiding a resurgence in market gardening. Perhaps the fastest-growing sector at present is that of organic production. At present the demand for produce grown without the use of chemicals in a way that cannot harm the environment far outstrips current production. This must be one option that any market gardener must give serious consideration to. With this development the history of market gardening has completed a full cycle by returning to its roots (no pun intended). To become a fully fledged organic grower may necessitate a total rethink of the way crops are grown and handled. It is something that should not be taken on half-heartedly but can be made to work. I know of one holding in the north of England that has been producing organically both outdoors and under glass for many years. They do not even ask a premium on the price of their produce. Nearer to my home in East Anglia there is a market garden which decided to convert to organics about three years ago. The grower concerned expected many problems, especially with pests and diseases but has found that his fears were totally unfounded. The demand for his produce has grown out of all proportion and he only wishes he had made the change earlier. Perhaps the biggest problems the modern organic movement have are its image and the fact that those involved cannot agree what exactly constitutes 'organic'.

A modern market garden.

A modern farm shop with imported and exotic fruits, etc.

Many modern market gardeners may fall in a sort of half-way house: not able to afford or want to use the vast array of modern chemicals but not totally committed to the organic movement. They do have certain marketing opportunities, especially freshness and the fact that they are local; these should be fully exploited to make the business successful.

Life has been made much easier recently. The range of machinery and equipment available now has enabled the modern market gardener to utilise his skills to the full. Sitting on a tractor or planting machine may at times be hard work, but does leave the grower with more time and energy to devote to other jobs which put the finesse on his product. The modern pedestrian-operated rotavator and the small or compact tractor are ideal tools; they can do a wide range of tasks and save time and grower's energy. There is also a wide range of other growing aids. The development of the modern floating films and poly-tunnels has reduced the amount of easily breakable glass on holdings. Gone are the days of glass cloches and the care that handling them involved. Now tunnels and rivers or lakes of floating film have replaced them.

The advances in plant nutrition and pest and disease control have all contributed to make the modern market garden. Developments in plant breeding and new and improved strains have increased yields beyond all recognition.

The modern market garden is not all muck and mystery, and gaiters and horses as we may think, but a modern highly efficient agro-business, producing a high-quality product to satisfy a changing local demand. We will now move on to see how this can be achieved.

2 Marketing

Marketing is often forgotten when a new market garden venture is set up. Yet it is the one main principle which can make or break the developing business. A well-devised marketing strategy can give the business a much greater chance of success right from the start; it is far better than forming one out of necessity over a number of years. This is not to say that the strategy should be inflexible. Having a definite strategy to start with allows one to change from a known position, taking all outside influences into account before changing the policy.

Before we go any further, let me make one point which I consider to be the most important single point for success. This is: grow to an outlet or outlets. Do not grow and then think 'Where can I sell this?' For those considering setting up a new business this may appear to be a very difficult thing to do but, like all aspects of a new business, good planning and forethought are worth their weight in good returns.

For example, imagine you have found what seems to be the ideal holding – cheap, south facing aspect, deep fertile soil, natural water supply, etc. – and it is 35 miles (56 km) from the nearest large town. Closer than that are perhaps three villages, four farm shops and another market garden. Having planned a good production for the first season of growing on your new 8-acre (3.2 ha) holding, you then have lots of lovely produce ready to sell, and then what? The local shops already have their suppliers, the nearest market is 35 miles (56 km) away. You line up a good outlet which wants three deliveries a week – not a large amount at first, but with potential. Now let's look at this more closely: with three trips per week to this outlet,

Three return trips, i.e. 3 × 70 miles = 210 miles (3 × 112 km = 336 km)
Vehicle returns 25 mpg = 8.4 gallons of fuel (9 km/l = 37.3 litres of fuel)

This works out at current prices at approximately £5.00 per trip, plus cost in time of a 70-mile (112 km) round trip, and the cost of operating a vehicle. At a very low estimate all this adds up to £20.00 per trip. What is the profit margin on the trip? If it is a small delivery it could be less than £20.00, so delivery or transport can obviously play a vital role in the profitability of the business.

Looking at the above example, what other possibilities are there?

Marketing

Local outlets? Are they big enough and can you break in on existing suppliers, a farm shop? There are four in the area already and probably only one good one but it is 35 miles (56 km) from a main centre of population: Are you on a main road? What other attractions are there in the area? There will possibly not be enough customers to rely on. Perhaps the 5-acre (2 ha) holding with a tattier house only 5 miles from the town may have been a better buy after all.

Is all this just hypothesis? I'm afraid not. The type of holding and its location to the proposed market is of paramount importance. Soil fertility can be improved, but costs of transport or, more generally, marketing costs, cannot – they always go up. Location is also of paramount importance when considering a farm-shop type of enterprise or a Pick Your Own. If the holding is not to be the main source of income, then the proximity to markets may not be of such importance, but it will still have a very substantial effect on the viability and profitability of the venture. If you have already purchased your holding this may influence considerably what and how you grow.

This is now an appropriate time to consider what the potential outlets are. Obviously they will relate to the types of crops you are growing or plan to grow. Basically, these outlets can be subdivided to on-farm or off-farm outlets. So let's now see what the possibilities are.

ON-FARM OUTLETS

Farm Shops

These can have many advantages with no transport, no packaging, everything under your own control, and the maximum price achieved for the products as no middlemen are involved. The disadvantages may include the wide range of crops that need to be grown; location (it will need to be near a centre of population or have good passing trade – preferably both), good road access and parking area are important; the local competition will also have to be considered. What are the planning requirements? (See Chapter 9, page 116.) Permission may be required if you sell produce not grown by you. Also, it may be difficult to get staff.

Farm shops are often associated with pick your own enterprises. These two types of outlet do seem to complement each other but is your holding big enough, and is it situated near a large centre of population?

A purpose-built farm shop.

Good sensible layout, but a shame about the boxes. An organic farm shop in East Anglia.

Marketing

Runner beans, an ideal Pick Your Own crop.

Pick Your Own

These enterprises lend themselves to soft fruit and a range of vegetables such as peas, beans, sweetcorn, courgettes, etc. As with farm shops, location will play a vital role in the success of the enterprise, as will the organisation of your own resources. A suitable cash point is required, good signposting, etc. Security should be high on your list of priorities, as should be a good car parking area. Do you need to consider other ancillary facilities, i.e. toilets, picnic area, children's play area, etc.?

Pick Your Own enterprises by their very nature have a seasonal bias and this can cause problems for a sole-income business with regard to the annual cash flow.

It may be possible with both of these types of enterprise to widen the range of products being offered and help to complement their seasonal nature by extending the season or filling in quiet times with pot plants, bedding plants and even nursery stock, especially container grown herbs, shrubs and trees. A number of modern garden centres now sell vegetables and fruit, so it may be time to take a leaf out of their book, so to speak, and sell plants as well.

OFF-FARM OUTLETS

Local Street Markets

Taking a stall in a street market still allows the grower a certain amount of control so it may be a useful outlet for produce. But several factors need to be considered: are the staff available to run the stall and are any required back at the holding? Will you be able to top up the stall on busy days or does distance and lack of staff make this impossible? As a second outlet this option may work well, enabling the grower to reduce surpluses and yet retain the maximum price for his product, for yet again, like both the farm shop and PYO, no middlemen are involved. Such a stall will also reduce the costs of packages as boxes can be reused. To work successfully it requires a reasonable range of products displayed regularly, which could cause problems with cropping schedules. Also, depending on the day of the week, it may mean a reduced volume of produce is available the following day. This could be a problem if the market happens to be on a Friday, and Saturday is your busy day in the farm shop. Which is it to be – is it more acceptable to run out on the stall, or in the shop? This would be an option worth investigating before finally settling on a marketing strategy.

Marketing

Local Retailer

Now we have to go a little further away from home to the local retailer. Here again regular supplies are required. It should be possible to build up a good personal relationship with the retailer and establishing good communication will help considerably. The volume of product taken may vary from retailer to retailer, and this may cause transport problems as in the example earlier in this chapter. However, assuming they are within economic distance, the local retailer may be a good outlet as it enables you to get on with the growing and keeps you away from the general public! Prices received from the local retailer should be above those quoted as Market Wholesale prices, and regular checking in trade magazines and local newspapers will ensure prices are realistic. The local retailer should also gain benefit by getting a fresher product from a locally known supplier. To achieve this, you the grower need to remember that in harvest quality, and grading, you are competing with the big boys, but you have freshness and adaptability (adjusting grading and crop requirements to retailers' requirements) on your side. With a little effort, a good working relationship can easily be achieved to the mutual benefit of both parties.

Catering Establishments

These still retain the local flavour, or may in fact be looking for local flavour. They may have general requirements for good-quality produce but could have very specific requirements, for example courgettes with the flowers still attached. There have been several notable cases of growers finding that the catering establishments in their area could not obtain a range of products and so set out to satisfy their demand. To do this successfully one has to work closely with the chefs concerned and pay great attention to detail in the crop husbandry. When involved in this specialist market a very high return is required for such specialised crops, although it may only be a case of leaving a few inches of stalk on the young carrots or some other such criterion which would normally cause the produce to be of an inferior grade according to statutory standards. The viability of such specialist growing relies on a sufficient size of demand, and being close to an area where there are several up-market hotels or restaurants will obviously help. Another possibility is for so-called exotic vegetables for ethnic minorities. These could range from peppers, through Chinese leaves to fresh spices. These could then be sold to restaurants and other ethnic food shops.

Local Wholesalers

These are sometimes referred to as secondary wholesalers but they can be considered as one and the same. Their main business is procuring produce to supply a large number of local retailers, restaurants and hotels with a full range of produce. They procure these supplies from local growers, and the major wholesale markets. They are, if you like, the ubiquitous middlemen. As such though, they can serve a very useful purpose.

These local wholesalers can be a major outlet for your produce as they can often handle relatively large volumes of product. Even so, to gain the best from them for both price and service, it is best to build up a good relationship with them. Dare I say, this is best done by good communication and a regular supply of quality product. If this local wholesaler knows he gets regular product from you he is much more likely to help if you have an unexpected glut of one crop. Also, if your quality is always high and he knows the bottom boxes are as good as the top, he is much more likely to take your product in times of glut in preference to another down the road who drops surplus product of indifferent quality on him occasionally. So once again, for the best returns it is consistency, quality and communication that achieve the best prices and service.

I have not yet said much in detail about communication, though I have mentioned it in passing several times. What do I mean by this? Well, it really ties in with the other two criteria, namely consistency and quality. If, for example, you normally supply him with fifty boxes of, say, lettuce on Thursday and something happened to the crop, such as mildew or the goat got out on Tuesday, ring him to tell him you can only supply twenty-five boxes on Wednesday or as soon as possible, rather than wait till he turns up to collect the produce. If this latter occurs, it may mean he has to let down his own customers, something he won't like doing and might take out on you! If you have let him know in good time he can make alternative arrangements and find other suppliers or even warn his customers. This all helps to build trust and mutual respect. The same is true with regard to quality as well. It also does not hurt to ask regularly about prices and other market trends, volume of supply, new products, etc.

Although the prices returned by such people are not necessarily the best, they have to make a living and run the costly transport. I would suggest that every grower should be on good terms with at least one local wholesaler – even if the volume he may actually take is only small, it can be worth every penny if the volume you have is too large for your other outlets to take but too small to deliver to a major market.

Marketing

If you do have a good relationship with a local wholesaler who makes regular calls to the major markets, he may deliver your produce there at a very reasonable price; his lorry may well be travelling to London or Birmingham empty to collect oranges or bananas and to get some money for the empty trips is obviously worth his while. So if you do want to get produce to the major wholesalers this may be a very economic way of doing so.

Major Regional Markets

These are markets such as Nine Elms (New Covent Garden), Spital-fields, Western International, and Stratford in London and others in cities such as Birmingham, Manchester, Leeds, Bristol, Cardiff, Newcastle, Glasgow, etc., around the rest of the country. The demands of these markets are similar to the local wholesaler as regards continuity, quality, etc., and the building of a good relationship may be even more important as almost all contact will be over the phone and the personal touch may be difficult to achieve.

They may also be very time-consuming to deliver produce to. Most markets open about 4 a.m. or earlier, with most sales occurring around 6 a.m., and they are relatively quiet by 8 a.m. So produce will need to be delivered in the early hours of the morning and you may not get home till late morning. Up at midnight, followed by a three-hour drive, humping boxes about, and a three-hour drive home is not inducive to a hard afternoon's work and if done regularly can be very exhausting. These are the markets that can handle large quantities of produce because of their sheer size, but all that has been said before about continuity of supply and quality is still of vital importance.

One other factor that now shows real importance, much more so than in the previous categories, is that of packaging. Packaging is not something that should be skimped on. A visit to any of the regional markets will soon show the value of good packaging. By means of the packaging you are trying to project an image to the final customer.

A good-quality box is a must. Not only does it look good, it will also protect your hard-won produce during the rigours of transport. Just stop to think for a minute how well *you* would have to be packed to be loaded onto a lorry, driven over bumpy roads, unloaded and put into store, banged about, reloaded onto another lorry, driven 250 miles (400 km), unloaded, etc., and your only protection is a cardboard box! Remember that fresh produce is a lot more delicate and susceptible to damage than you are, so don't skimp. Also, remember to pack produce into appropriate boxes. No *English* produce should go into *Dutch* tomato trays, and banana boxes are for bananas, not lettuce or carrots.

18

The one problem with good-quality packaging is the price. If you are only producing relatively small quantities of a range of products it may not be possible to develop your own packaging and logo to the upper standards. You are then limited to the materials you can purchase from your local supplier. Another option open to you may be to group together for marketing purposes with other local growers to enable you to gain the benefits of bulk purchase and also possibly transport and other benefits. I discuss the benefits of co-operation overleaf.

National Marketing Agents

The Geest Organisation, Mack, and the Hunter-Saphire Group are three examples of these organisations. These are basically large commercial companies which obtain produce from both their own production units and from a range of other growers. The arrangements vary from a service similar to the local wholesaler to contracts for produce. All of these may enable you to reach outlets otherwise beyond the reach of the smaller grower, namely the major UK supermarkets. The supply of produce to these prestigious outlets can be fraught with difficulties but the rewards can be good. The standards laid down for supplies of product may be higher than those normally required by the Ministry or EEC Standards, the times of deliveries can be very narrow and the quantities required may be large but the returns can be good, or at least more consistent than market prices. This latter point is often argued but it can be proved time and time again that, taken over a season, the supermarkets return the best prices per unit of produce.

These prices can be bettered by the smaller grower if he or she specialises in out of season production, in which case the wholesale markets will almost certainly return the best prices. This premise points to another major consideration for the smaller grower – that of specialisation in either crops or timing of crops. One small word of caution with regard to specialisation: ensure there is a real market for your specialist crop: the courgettes with flowers still attached, as previously mentioned, may fetch a high price with local caterers but there may be no alternative outlet. The markets may only treat them as very small and unwanted courgettes.

If after all this you would still like to market to the national supermarkets there may be another way, via one of the National Marketing Co-Operatives such as Gloucester Marketing Services or Cambridge Salad Producers. Once again, volume and geographic location may prohibit joining one of these organisations as a member. It may still be possible to join a local marketing co-operative. We now need to look at these in more detail.

Co-Operatives

These can offer many advantages to the smaller grower. By working together with other like-minded growers marketing costs can be reduced, particularly transport and, as mentioned previously, packaging costs. The provision of a centralised packhouse can enable specialist grading and processing equipment to be purchased and used, as the cost is spread out over a number of growers. This in turn gives a better-quality product to enable the co-operative members to compete with the larger suppliers on more equal terms. Co-operation can then lead to other advantages for the smaller grower. These can include the passing on of discounts for bulk purchases of other requisites such as fertilisers, chemicals, etc. – even machinery pools can be formed to allow the purchase and use of specialised machinery. The machinery may not need to be that specialised to be advantageous. A polythene-laying machine may be above the reach of an individual grower but if several growers or the co-operative purchase it, the cost is spread and the growers can benefit from the use of such equipment.

If there is no co-operative in your local area which you could join, it may be possible to form one. Grants may be available and there are several organisations that help people to set up co-operatives, both from marketing and other production standpoints.

What are the disadvantages? The main one appears to be the regulations which have to be provided when a number of people own and run (via a board of directors) the Company. These are set to ensure the fair use of the company and that all members are treated equally. At times co-operatives seem to cut across out native British trait of individuality. But in practice most co-operatives work very well for the mutual benefit of all their members, as each member has a real say in how their organisation is run, unlike some of the other organisations I have already mentioned. This may suit some people and not others. In the end the final choice is yours, and should be that form of marketing which is most apt in your own situation.

To summarise then, the points to remember are as follows:

1. Grow for a specific market – that which suits your requirements.
2. Only grow for those markets you can conveniently reach, or select another site.
3. As far as possible, spread your market outlets so you do not rely on a single outlet. If your local retailer goes bust or unexpectedly stops trading can you cope? To use the age-old expression, don't put all your eggs in one basket.

4. Consider all the options before making any decisions.

All the points raised in this chapter are equally relevant whether the produce is chemically or organically produced.

3 How to Grow – Soils and their Management

Having looked at the vagaries of marketing in the previous chapter we now come back down to earth – literally. It is now time to consider the major resource of your holding other than the human factor, and that is the soil. The maintenance of a fertile soil should be the prime consideration of anyone lucky enough to be a custodian of land, be they either owner or tenant. This major resource, on which many a nation's prosperity has been built, has a vast influence on your own profitability. It is a fragile ecosystem of its own which can be easily damaged but can be improved and at times vastly improved. It needs to be carefully understood and treated with respect.

A good soil makes life easier, a poor soil makes the struggle so much harder, but take heart those of you that have a poor soil – you can improve it. However, there is no quick route, just good husbandry and understanding of this most basic of resources. 'The answer lies in the soil' is an old adage but there was never a truer one. The answer to many growing problems lies in good healthy plant, and that needs a good, fertile, well-nurtured soil. So how do we achieve this ideal soil?

The obvious answer to this question is to ensure you buy or rent good soil with an advantageous aspect in the first place. These should be the prime considerations; but, thinking back to the previous chapter, it is no good finding that ideal site – a good, fertile soil on a southerly aspect – if there is no outlet for the resultant mountain of produce. So when making an initial choice, find the best site within a reasonable distance of the market you want to grow for.

We have to look at the soil and see what constitutes the best soil. As it is so unlikely that we will find the 'ideal' soil, perhaps it would be better to see what soil is, to look at its structure and see what can be done to improve it. So what is soil? According to *The Concise Oxford Dictionary* soil is an 'upper layer of earth in which plants grow, consisting of disintegrated rock usually with an admixture of organic remains, mould.'

For all practical purposes, soil is a medium into which plant roots foray searching for the nutrients to sustain the growth of the plant. It also enables the plant to support itself to gain maximum benefit from

22

the available light. So the soil needs to be compact enough to support the plant, yet porous enough to allow the roots to foray easily to obtain the foods the plant requires. It is a hard compromise to achieve, and the level of difficulty depends on the properties of the main constituents of your soil. These are sand, silt and clay. A fourth constituent which can have a major effect on the soil structure is organic matter. It is the varying properties of sand, silt and clay which determine the classification of the soil (*see* Fig 1).

SOIL ANALYSIS

One rough and ready way of determining your soil is to half fill a jam jar or other suitable glass container with soil, add water until it is about three-quarters full, add some vinegar, put a lid on and give it a really

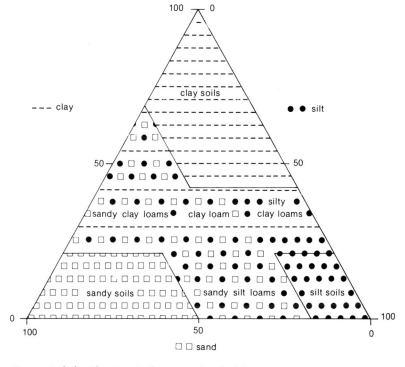

Fig 1   Soil classification. Soils can be classified by estimating the percentage at each fraction on the diagram.

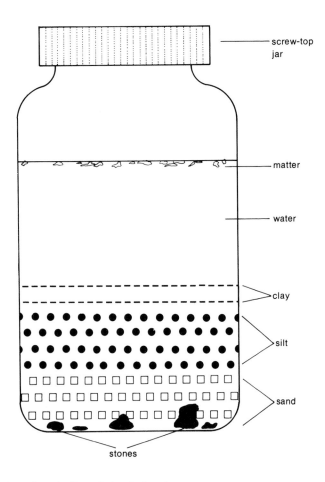

Fig 2 *Separation of sand, silt and clay in jam jar.*
Approximate proportions of 50% silt, 40% sand, 10%
clay give a sandy silt loam.

good shake. Shake for about one minute and then put it down imme-
diately. The vinegar helps to settle the fine clay particles quickly. The
various fractions of sand, silt and clay will separate and an approximate
proportion of each can be ascertained.

The sand and any stones will be at the bottom, followed by the silt,
with the clay on top. Any organic matter will be either floating on the
surface or on top of the clay. From this you can check quickly on the soil
type of your holding. Remember, though, that soil types can vary
considerably in a small area, and there may be considerable variations

across your land, especially if it is on a slope. This can influence decisions on where to site various parts of the holding, such as glasshouses, buildings, etc.

A simpler method of assessing soil texture is to rub a sample of moist soil between the thumb and fingers. A sandy soil will feel gritty, a silt soil will feel smooth and silky, and a clay soil will feel sticky. Why do we need to know these properties? The workability and base fertility of the soil depend on this basic make-up, and if we look at the main ingredients in turn we can see why.

Sand

Sand, by international convention, is defined as soil particles which vary in size from 2mm to 0.02mm in diameter. They are made up of solid, insoluble particles of base rock, which greatly improve the aeration and drainage of the soil but provide no real amount of nutrients for the plant. A soil with a reasonable proportion of sand will be easily workable, will not become waterlogged, and will warm up quickly. However, if the soil has a high proportion of sand it will be a dry, hot, hungry soil, perhaps best used for early production especially if with a southerly aspect.

Silt

This is the next size of soil particle, ranging from 0.02 to 0.002mm in diameter. Although particles of silt are much smaller than sand particles they are still basically unweathered minerals and therefore contribute only a little to plant nutrition. Their presence will increase the water holding capacity, but they can impede drainage by blocking up pore spaces; yet they will give the soil more bite.

Clay

Clay particles are the smallest, being less than 0.002mm in diameter, and are very different from sand and silt particles. They are the products of chemical weathering of other soil constituents. These particles are extremely small and are in fact of colloidal size; because of this they have a large surface area in relation to their mass. This means they have many surface atoms and, as they are electrically charged, they can attract ions from the soil solution and hold them on their surface. Usually they are negatively charged and attract the positively charged ions or cations, e.g. potassium, calcium, etc. Therefore they are the major influence on the fertility of the soil.

It can be seen that the fertility of a soil is in a direct relationship to the proportion of clay particles within the soil, all other aspects being equal. This is fine, but from the grower's point of view the workability of the soil has an inverse relationship to the amount of clay, i.e. the greater the proportion of clay the heavier the soil, and therefore the more difficult to work it becomes. Once again, the best soils are in the middle range, providing an ease of workability yet having a sufficiently large proportion of clay to be fertile and provide a buffer reservoir of nutrients for optimum plant growth. Whilst I have not covered this area in great depth, it should be enough to enable the reader to have sufficient understanding to ensure they are aware of the potential of various soil types.

We have now looked at the basic make-up of the soil; now we must consider it in more depth – literally – and look at the soil profile.

SOIL PROFILE

The soil profile can have a profound effect on any individual soil type. It is not just the top 9 inches (23 cm) of soil with which we must be concerned but also what is underneath. A soil profile is a vertical section through the soil and to gain a full insight needs to extend at least 3 feet (91 cm) down and preferably more, to see what is below. For this purpose the soil profile is divided into distinct layers or horizons which may or may not be easily recognised. There are usually considered to be three horizons: the A horizon, or topsoil, the B horizon, or subsoil, and the C horizon, or parent material.

The depths of these individual horizons can vary considerably. The A horizon or topsoil, can vary from just a few inches to several feet but is normally 9 to 12 inches (23–30 cm) as this is the normal depth of cultivation.

When assessing a site, the subsoil, or B horizon, can be of great importance. It may be a heavy clay which will impede drainage and could have an adverse effect on the topsoil, or it could be a free-draining subsoil overlaid by a clay topsoil. Many combinations can be found; and being aware of what you have, and how this influences cultivations and plant growth, can make your life much easier. This may prevent problems occurring, or help provide reasons why certain things have been happening.

An impervious clay layer or a hard 'pan' in the subsoil will restrict drainage and plant root growth, and hence reduce the cropping potential of the land. It may be possible to help the situation by careful management or the use of mechanical equipment such as subsoilers to reduce these problems.

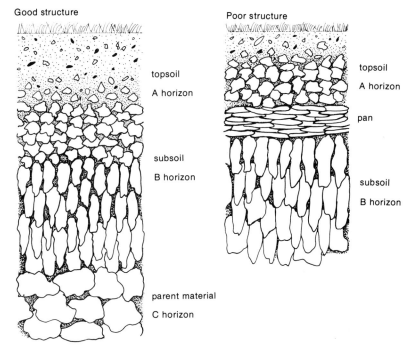

Good structure

topsoil
A horizon

subsoil
B horizon

parent material
C horizon

Poor structure

topsoil
A horizon

pan

subsoil
B horizon

*Fig 3   Soil profiles, showing good and poor structure,
horizons and a pan.*

One convenient way of finding out the basic soil types is to look at the maps of the Soil Survey of England and Wales which should be available at the local reference library. Whilst these may not be totally accurate for your land they will certainly give a good indication of what you can expect and, if you are looking for a suitable site, may narrow down the areas to be examined in more detail.

As we have seen, soil texture and soil profile can have a great influence on the cropping potential of an individual site. Those areas that have a good soil are usually the most expensive to buy, and with good reason. It is very difficult to change the soil texture appreciably. It can be done but will be a long-term project, the full fruits of which may not be seen in the lifetime of those starting the process. Having said that, it is no reason not to plan into the management of the holding all the techniques that will help to achieve this. At the very least, this will prevent a deterioration of this most basic of the grower's or mankind's resources. This form of conservation or improvement is actively encouraged by the current Organic movement as a vital part of any growing or farming system.

It must be remembered that soil texture is, if you like, the mix of the basic ingredients of the soil, in their physical proportions. Obviously, to change this physical proportion, vast amounts of sand or clay would need to be added. Take heart though, especially those of you who have a heavy soil, there is one area we can change relatively easily, and that is soil structure.

SOIL STRUCTURE

The analogy of a house may help to explain what soil structure is. If we assume the wood, bricks and mortar to be sand, silt and clay (soil texture), they can be put together in the same proportions to construct a house, but the actual style of house can vary considerably using the same amounts of wood, brick and mortar. The same is true with soils and this arrangement of the soil particles is known as its structure.

A good soil structure will allow water to enter the soil easily and excess moisture to drain freely from it. It also allows air to permeate the soil, as without air plant roots cannot live. With a good crumb structure cultivations and the preparation of a good tilth will be easy, and the soil is less likely to be damaged.

What do we need to obtain a good crumb structure? Ideally the soil crumbs should be in the size range of 0.5 to 5 mm. The soil's ability to form a good crumb structure depends on the amounts of two constituents, namely, clay and humus, or organic matter. Now I have already mentioned that clay particles help to bind the soil with larger particles but, as we have seen, if the proportion of clay is too high the soil will be impervious. Humus or organic matter will also have the same effect of binding the particles together to form crumbs. It is here that we have the greatest opportunity to improve the soil. We all know that we should add compost or other organic matter to a soil but not necessarily why. It will improve a sandy soil by helping to bind the particles to stabilise the soil and make it more water retentive and with a clay soil it helps to separate the clay particles into crumbs and make the soil more porous. The addition of organic matter will therefore help the structure of a soil immensely. It will also give other benefits which we will discuss a little later.

A good soil structure will provide the best conditions for plant growth. To be really effective it needs to continue as far as possible down the soil profile, and ideally throughout the subsoil or B horizon. It obviously becomes more difficult to modify the soil structure the further down the profile one goes, but steps can be taken to do this which are covered in the following section on Soil Management. Before

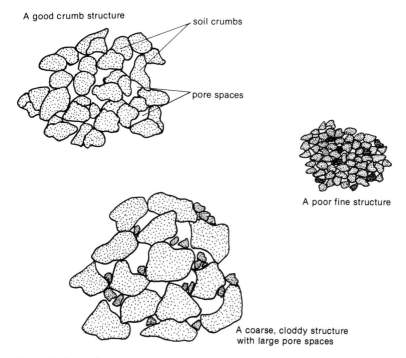

A good crumb structure

soil crumbs

pore spaces

A poor fine structure

A coarse, cloddy structure
with large pore spaces

Fig 4   Soil crumb structure.

moving on to soil management it would be best to summarise briefly.

Soil texture identifies the proportions of the major building blocks of the soil, i.e. sand, silt and clay, within the soil profile. The soil structure is the way these particles are actually arranged. It is this latter that we can influence most, and to provide good crop growth a good structure is essential. It is the soil structure that we, as custodians of the land, can most easily improve or destroy depending on how we manage our soil.

SOIL MANAGEMENT

The most important aspect of soil management is timing of the various operations. Anyone who has worked the soil for pleasure will know the difference of working a soil at the correct time, especially on a clay soil. If it is too wet it can be sticky, while a few drying days later it is like concrete and nothing can break up the lumps. It is bad enough trying to time it correctly on a small vegetable patch but when the pressures of running a commercial holding are applied to the situation, it can be very

Rotavating a light sandy soil in good conditions.

difficult to say, 'It's a bit wet today – I'll leave the rotavating till the soil is a bit drier.' This may be fine if time permits, but if the plants have been ready for planting for a week already and there is a bad weather forecast, what alternative does a grower have, except to do the job at a less than ideal time? This incorrect timing of cultivation can at the least provide the plant or seed with inferior conditions, and at the worst drastically damage the soil structure. Hopefully any damage to the soil structure will be limited, but it can be virtually permanent. For example, lasting damage was done to many fields in East Anglia where sugar beet was harvested in wet conditions in autumn 1987. In many cases, the damage to fields incurred at the access points is irreparable.

Damage to the soil can be done in two ways. First of all, the smearing action of a lot of cultivation equipment, especially ploughs and rotary cultivators, can cause major problems of panning, especially if cultivation is always done at a constant depth. Certain soils are more susceptible to cultivation pans than others, especially those with a high clay content. Pans are created by the smearing action of the machinery which makes the soil below impervious, and so impedes drainage and the passage of air, both so important in the root zone. They can be prevented by varying the depth of cultivations, ensuring the soil is in a suitable condition for cultivation, and also by the use of subsoilers. To gain maximum benefit from subsoiling this operation needs to be done when the lower levels of the soil are relatively dry. This ensures much greater fracturing as the blade of the machine is pulled through the soil. There is also a second advantage of timing this operation to coincide with a drier soil. As the leg of a subsoiler extends some 2 to 3 feet (61–91 cm) down, it requires a fair amount of power from any tractor pulling it. This tractor will do less damage to the soil the drier it is. There is no point in using a subsoiler to alleviate the problem of a pan in a soil, yet increase compaction and smearing action in the topsoil by doing so.

The second way in which soil structure can be damaged is by compaction of the soil. A good soil with a balance of the main ingredients of sand, silt, clay and organic matter will have a much better resistance to compaction than either a clay soil or a sandy soil, or in fact any soil lacking in organic matter. This is due to the ratio of pore spaces within the soil. The passage of machinery over the soil will have the effect of closing up the large pore spaces. The weight of the equipment pressing downwards pushes the individual soil particles closer together. This reduces the amount of pore space and hence the speed with which water can be both absorbed into the soil and drained away. It will obviously affect the amount of air in the soil as well. Both these factors can in turn limit the growth potential of any crop in such a soil. The

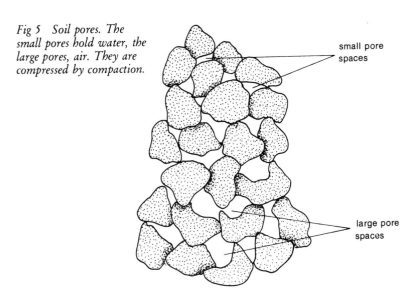

Fig 5   Soil pores. The small pores hold water, the large pores, air. They are compressed by compaction.

small pore spaces

large pore spaces

repeated use of rotavators on clay soils, continued with a lack of regular dressings of organic matter, will also destroy the structure of the soil, even with correct timing of operations.

We have now seen that working the soil can cause real damage to the structure and fertility of the soil. What can we as growers do to ensure that damage is minimised and the structure maintained or improved?

Firstly, the easiest way any tiller of soil can ensure that he or she does minimal damage is by timing all operations to coincide with the correct soil moisture content. A sandy soil can be broken down to a fine tilth when dry but with a clay soil timing is critical; too dry and the lumps are too hard, too wet and it's like modelling clay, and the cultivations will do further damage to the structure.

Secondly, the depth and type of cultivation equipment used should be varied, if indeed any form of cultivation is used. The specific machine should be matched to soil type and moisture content to get optimum results. Next, use equipment that has a low ground pressure and make as few passes as possible. It has been estimated that in normal spring cultivations and sowing approximately 90 per cent of the soil surface is covered by wheelings at one time or another. To put it another way, the potential for compaction is 90 per cent of the field! This may lead to the conclusion that it would be best for organic and commercial farmers to use a no cultivation system, but this would only be so if no equipment or wheels pass over the land.

How to Grow – Soils and their Management

The third way we can maintain and even improve the soil structure is by the addition of organic matter. This will have many effects: it will improve structure, increase water holding capacity and nutrient supply (especially nitrogen), and make the soil more resistant to compaction. Such improvements will make the soil much easier to manage, and the timing of cultivations will be less crucial.

There is a variety of ways in which organic matter can be added to a soil. Sufficient organic matter will exist in pasture-land soil which is newly ploughed. It has long been known that the soil is vastly better when a permanent pasture is cultivated. This effect only lasts for a few seasons, and further organic matter must be incorporated with subsequent cultivations. On natural soils the residual plant material from the native vegetation dies and returns to the soil, and a slow increase in organic matter occurs in the topsoil. In an intensively cropped system the waste plant material is often removed from the field. This may be done to help prevent a build-up of pest and disease organisms on the site. On a market garden no resource should be wasted so any surplus plant material should be returned to the soil to help maintain its structure and fertility. To prevent the spread of pests and disease organisms, the material should be composted.

When considering the long-term effects on soil of intensive cropping it must be remembered that a large proportion of the plant material grown on the site will not be returned to the soil as it is the marketable product. This needs to be replaced in some other form. Farmyard manure is the most common form of organic matter to be added to soils. This can be added direct, or composted with the holding waste and then spread on the land. But it should be remembered that farmyard manure is a variable product and the amount of nutrients it will supply to a soil can be variable.

Straw is another organic material that may be readily available in certain parts of the country. Some words of caution are needed though if fresh straw is added to a soil. Because of its low protein content (proteins are required by soil bacteria to break down the material), it will actually give a reduction in nitrogen levels for the succeeding crop. The addition of $4\frac{1}{2}$ tons of straw per acre (12 tonnes/ha) will in fact lock up $62\frac{1}{2}$ lb of nitrogen per acre (70 kg/ha) from that available for a crop.

Another useful way of adding organic matter to soils is by green manuring. This is the practice of growing a 'crop' and incorporating all of it directly into the topsoil. It will be no stranger to the devotees of organic growing. In many areas it can aid soil conservation and soil improvement.

One major factor that has led to the problems of soil erosion in some areas is that of leaving a soil uncovered over winter. This can lead to

Straw being composted in an organic market garden.

*Pedestrian rotavator with rake attached for seed-bed
preparation. A relatively cheap machine.*

topsoil being washed off sloping land by winter rains. The use of green
manures will stabilise the soil during this time and give a benefit of
additional organic matter when incorporated in spring. In fact, green
manures can be beneficial at any time when a soil is not to be cropped for
a while. Not only will they stabilise the soil, they can help retain
moisture and suppress weeds, and, depending on the plants used, help
to draw up nutrients from the subsoil as well. This latter will occur
especially if deep-rooted types such as lucerne are used.

A good rotation of crops also plays an important part in maintaining
soil fertility. As crops have differing needs, the soil is given a chance to
recover, which is not the case in a monocrop situation. Where possible,
some form of rotation should be used. Having said that, it could be that
specialisation in cropping may be necessary to ensure the economic
success of a modern market garden. If this is so, real effort should be
made to prevent the possible harmful effects of monocropping. The use
of farmyard manure, compost or green manures will help restore a more
natural balance.

REASONS FOR CULTIVATION

I have examined the importance of timing of cultivation with regard to the possible damage to soils. In many respects, if we reflect on the previous section, a non-cultivation system may seem appropriate – so why do we cultivate the soil? The basic answer to this question is to provide suitable conditions for root development for the plant.

We have seen how compaction and naturally formed 'pans' can hinder root development, impede drainage and generally make life difficult for the plant. Therefore, the aim of any cultivation must be to improve conditions for the plant. It will also aid the incorporation of organic matter and fertilisers. There is now a vast range of cultivation equipment available to help the modern market gardener, and great care must be exercised when choosing equipment – will it do the job I want, is it the best type of machine for my growing system, my soil type and my budget? Cultivation equipment will be discussed in Chapter 6 (page 78).

4 Crop Requirements or Nutrition

In the previous chapter I examined the soil and recommended ways to provide optimum conditions for the crop roots to develop. We now need to consider the crops' nutritional requirements.

A crop will only reach its full potential for a grower if it is supplied with the right amounts of the basic foods required. This will of course vary from crop to crop, but before anyone goes out and spreads liberal amounts of fertilisers to promote growth, we should ascertain what is required and how much. So where do we start?

Chemical fertilisers are expensive, and not permitted if you are an organic grower, so it is always a good idea to have your soil analysed for its nutritional status. This can be done, at a cost by ADAS or other laboratories, or even by fertiliser manufacturers. Kits can be bought for DIY analysis but these will not be as accurate. They will, however, give a good indication of nutritional status and should register any changes in this status.

Of the three main plant foods – nitrogen, phosphorus and potassium – nitrogen cannot easily be analysed and has to be estimated from past cropping.

It will now be beneficial to consider each of the plant nutrients in more detail.

NITROGEN (N)

Nitrogen is probably the most important plant food. This is because it is very difficult to determine the optimum amount to provide for a crop. If there is too little the crop won't grow: too much, and maturity may be delayed and the fertiliser is wasted or even the environment damaged. The soil's natural reserve of nitrogen is contained in the organic matter within the soil. In mineral soils the vast majority of organic matter is in the topsoil so that is where the naturally available nitrogen will be found.

This organic matter is continuously being broken down by soil bacteria and converted to a form which plants can readily absorb. This

Fig 6 Distribution of organic matter in the soil profile, with normally about 5% organic matter in topsoil, decreasing with depth.

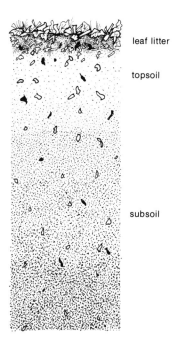

leaf litter

topsoil

subsoil

process has two stages: proteins in the organic matter are first converted to ammonia, which is then converted to nitrates by the nitrifying bacteria in the soil for the plants to take up. This all occurs happily when oxygen or air is present. If the soil becomes waterlogged and no oxygen is present, another range of bacteria attacks the nitrates and releases nitrogen gas. The importance of having a well-drained soil with a good structure can be seen once again. We can apply artificial fertilisers to make up the difference between the crop's requirements and the amount of nitrogen naturally available in the soil. The big question is how much to apply. As nitrates are readily soluble, the timing and amounts applied are critical, if any excess is not to be leached out into drainage water before the plants can absorb it.

As nitrates are so highly soluble and easily move through the soil in the soil water it is very difficult to assess accurately the nitrogen levels in the soil. The normal system used relates to the organic matter content of the soil and therefore relates directly to previous cropping of the area of land in question. An arable growing system with no pasture included in the rotation will produce an organic matter content of 1 to 3 per cent, as opposed to that of a ploughed-up ley where levels of organic matter may reach 5 to 10 per cent.

As we have seen, nitrogen levels relate directly to organic matter

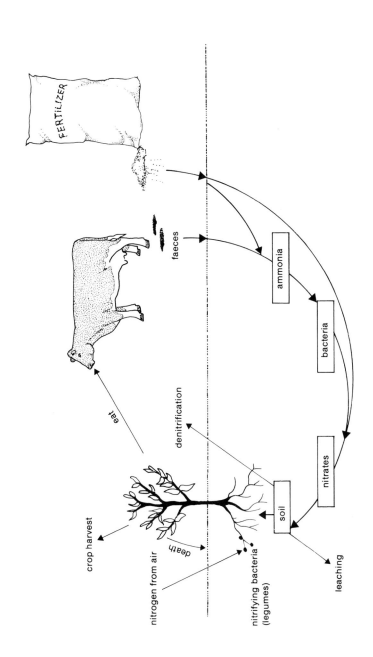

Fig 7 The nitrogen cycle. Under natural conditions plant nutrients are recycled. When a crop is harvested and removed this can lead to a lowering of nitrogen and organic matter levels.

levels; hence an intensive arable system will produce only a little natural nitrogen and will require more artificial fertiliser. The organic content of the soil should be increased by incorporating bulky organics, such as farmyard manure or compost, using a rotation including leys, and applying green manures, especially those utilising leguminous plants such as lupins or clovers which fix gaseous nitrogen in their root nodules.

PHOSPHORUS (P)

Phosphorus is vital to the plant for respiration and in the actively growing parts such as roots and shoots. It is also required in the seed for germination. Unlike nitrogen it is not leached from the soil. This is because the inorganic phosphorus becomes insoluble in the soil and accumulates from repeated fertiliser dressings. Phosphate deficiency is therefore relatively uncommon, especially on the heavier clay soils of the Midlands and East Anglia. Light soils, and the acid soils of Wales and northern England, are more likely to require regular dressings as the phosphates are rapidly converted to an unavailable form. Yet, once the phosphate has been absorbed by the plant it remains mobile and can be moved from older parts to the actively growing areas.

As crops such as potatoes, brassicas and roots have a higher demand for phosphates it may be possible to rotate the application of phosphate so it is applied only before a high-demand crop. It should also be noted

Brassicas have a high demand for phosphate.

that to obtain maximum yields one has to apply more phosphate than the crop will actually take up. It should be placed near the roots of the crop as it is immobile in the soil; the roots have to forage for it, so a good soil structure will help this process. Compacted or waterlogged soils will deny plants phosphates. If artificial media (loamless composts) are used for plant raising, ensure sufficient phosphate is added as there will be no natural reserves in the compost. Phosphate can be leached out of soil-less composts, so the use of slow-release forms is recommended.

Bonemeal is the traditional fertiliser but this is now believed to be unsuitable; natural rock phosphates should be used, or super or triple super-phosphates.

POTASSIUM (K OR POTASH)

Potash is generally considered to be the food for flowers and fruit. It is also used in the plant as an osmotic regulator and helps to keep it healthy, and plays a part in the control of photosynthesis and respiration. When in the correct balance with nitrogen, it will increase the plant's resistance to chill damage, drought and disease. These latter points can be of great value to the grower. The ideal balance to achieve is a 1:1 ratio of nitrogen to potassium within the plant. However, plants are capable of taking up more potassium than they actually need (luxury levels) and this can have an effect on soil reserves, particularly

Tomatoes may require heavy dressings of potash, but in balance.

where leafy crops are harvested young. For this to occur, high levels of nitrogen need to be applied.

The balance of potash with another plant food, magnesium, is also important, especially on crops such as tomatoes. Applying heavy dressings of potash can upset this balance and lead to a deficiency of magnesium. For this reason, when applying heavy dressing, such as for cherry tomatoes, part of the application should be made as magnesium sulphate (Kieserite) to maintain the balance in its correct ratio of 9:1 potassium to magnesium. Potassium is readily recycled from organic matter and is present in large amounts in young soils, but due to its solubility may be leached from old soils. It is also easily leached from soil-less composts and sandy soils with a low organic matter content. In horticulture the most commonly used fertilisers are potassium sulphate (sulphate of potash) and potassium nitrate. The latter also supplies nitrogen and is commonly used in liquid feeds; it is hygroscopic (absorbs moisture from the air) and needs to be kept dry.

MAGNESIUM (Mg)

Magnesium deficiency is probably one of the most common deficiencies to occur in intensive horticulture. It is most likely to exist on light soils with a low organic matter content or where heavy dressings of potash have been applied (*see* Potassium above). Magnesium deficiency can also be a problem on lime-rich soils. To counteract it, the most common fertilisers used are magnesian lime, which raises the pH as well as adding magnesium (for pH *see* Calcium below), or Kieserite (magnesium sulphate). If a deficiency occurs in a growing crop spraying with magnesium sulphate (the pure form of which is Epsom salts) will have an immediate effect.

CALCIUM (Ca OR LIME)

Calcium needs to be considered in two ways, firstly as a plant food where a deficiency shows in a variety of symptoms (*see* Table 2, pages 46–7). The second way calcium needs to be understood is in its most commonly applied form – lime. In the past lime was always considered to be vital to 'keep the soil sweet'. As with most old sayings, this bears more than a grain of truth. Lime is the major soil conditioner. It is used to correct the pH of the soil as well as supply calcium as a plant food. Generally speaking if the pH is correct there is very unlikely to be a problem of availability of calcium as a food. So what is pH?

The pH scale is a measurement of the acidity or alkalinity of a soil. It is in fact a measurement of the hydrogen ion concentration in the soil. To enable this to be expressed in simple terms, a negative logarithmic scale is used. This gives a scale from 1 to 14, 1 being acid, 14 alkaline and 7 the neutral point. In practice, the pH of soils generally lies between pH4 and pH8.5. Soil pH can be tested using either a BDH test kit or a pH meter. A pH test normally forms part of any soil analysis you may have done. It is very worthwhile to have a pH test done regularly to ensure the ideal pH is being maintained for the crops grown. The chemically neutral point, i.e. pH7, is not necessarily ideal for all crops and it is more usual to aim for a pH of 6.5 in most soils.

A soil with a low pH, i.e. acid, can easily have its pH raised nearer to a neutral level by the addition of lime. Before applying lime, two important points need to be remembered. Firstly, as the pH scale is logarithmic, a pH of 5 is 10 times more acid than a pH of 6 which is in turn ten times more acid than a pH of 7. Putting this another way, a pH of 5 is one hundred times more acid than a pH of 7. This will have an effect on the amount of lime required to raise the pH to optimal levels. The second important point to remember when applying lime is that it works only slowly in the soil, so its full effect of neutralising an acid soil is not complete for at least three years. If this point is not remembered it is very easy to overlime a soil. The potential of overliming or raising the pH too high is important as it is much easier to raise a soil's pH than reduce it. Lime also has a secondary effect on soils. It can actually help to improve soil structure, particularly on clay soils. One word of caution though: many clay soils, especially in East Anglia, have a relatively high pH, so check before using lime as a soil conditioner.

The main agents which act to reduce a soil's pH are acid rain (most of our rain is now polluted to a greater or lesser degree, but *see* Sulphur), farmyard manure and nitrogen fertilisers. Continuous cropping will also help to acidify a soil by plants removing calcium from the soil. The normally recommended agent to use to actively reduce a pH is sulphur (flowers of sulphur) which is expensive and difficult to apply.

The pH of a soil also has an effect on the availability of nutrients in the soil. Too high a pH can 'lock up' certain nutrients making them unavailable to the plant, i.e. lime-induced chlorosis or iron deficiency. Commonly, too low a pH can have a similar effect (*see* Table 2, page 46).

Crops also vary in their tolerance of pH. Most gardeners know rhododendrons like an acid soil and most brassicas like lime. A range of crops and their pH tolerances are given in the following table.

TABLE 1   pH TOLERANCES OF CROPS

	Crop	pH (All figures approximate)
Vegetables	Beans	6.0–7.5
	Brassicas	6.5–7.5
	Carrots	5.5–7.0
	Celery	6.3–7.0
	Courgettes	5.5–7.0
	Garlic	5.5–7.5
	Leeks	6.0–8.0
	Lettuces	6.0–7.0
	Peppers	5.8–7.0
	Parsnips	5.5–7.5
	Potatoes	4.5–6.0
	Swedes	5.5–7.0
	Tomatoes	5.5–7.5
Fruit	Apples	5.5–6.5
	Blackberries	5.0–6.0
	Black Currants	6.0–8.0
	Gooseberries	5.0–6.5
	Grapes	6.0–7.0
	Pears	6.0–7.5
	Raspberries	5.0–6.5
	Strawberries	5.3–7.5

See Appendix 1 *re* liming materials.

SULPHUR (S OR SULPHATES)

Sulphur is a vital ingredient in the synthesis of chlorophyll within the plant. The deficiency therefore shows as a paling of the young leaves of the plant. Sulphur is not readily transported round the plant due to its insolubility. It is normally taken up in the sulphate form and in the past has not normally been added as a specific fertiliser, but has been supplied either from the soil's natural reserves, from recycling plant remains (organic matter) or as a sulphate formulation of other fertilisers. A major source of sulphur has also been from polluted air where sulphur dioxide has dissolved in rain to form sulphuric acid, i.e. acid rain. With the reduction in the use of sulphates in fertiliser formulation and the non-recycling of organic matter combined with the modern clean-air policy, deficiencies may become more commonplace. In fact, it is now starting to appear in certain parts of the country, and has basically been put down to the clean-air policy!

TRACE ELEMENTS

We now come to the group of nutrients known as trace elements. This group of nutrients, though vital to the plant's growth and general health are required in small quantities, usually measured in parts per million (PPM). For this reason they are sometimes called micronutrients. Deficiencies of all these nutrients, with the possible exception of zinc, occur within the British Isles. The locations of these potential deficiencies depend on soil type and pH (*see* Table 3).

Manganese (Mn)

Manganese deficiency usually occurs on soils with a high pH, particularly if the high pH is due to overliming. It is also more prevalent on soils with a high organic matter content. As the pH drops, Manganese becomes progressively more available to the plant, even to the extent that it can become toxic. This is another good reason for ensuring the pH of your soil is correct. Any deficiency can be corrected by a foliar spray of manganese sulphate at a rate of 1 oz per 4 gal. per 9.5 sq.yds. (1.5 g/litre/2 sq.m).

Iron (Fe)

Iron deficiency usually occurs on lime-rich soils, hence its common name of 'lime-induced chlorosis'. It affects the young leaves first. The problem can be overcome by reducing applications of lime and phosphates and by applying iron chelates.

Copper (Cu)

This normally only occurs on light sandy soils or organic soils overlying chalk. Plants suffering from it tend to go very dark. It can also be a problem with livestock. Deficiency can be rectified by copper sulphate sprays.

TABLE 2 MAJOR NUTRIENT DEFICIENCY

	Nitrogen	Phosphorus	Potassium	Magnesium	Calcium	Sulphur
Deficiency most likely to occur	Intensively cropped land and low organic matter	Heavy clays, light sands and acid soils	Intensively cropped land	Lime-rich soils and low organic matter. Associated with heavy dressing	Acid soils	Only now showing deficiency
Veinal Chlorosis						
Intervenial Chlorosis			*	*		
Darkening of older leaves		*				*
Pale juvenile leaves						
Necrotic area on leaves		*	*	*		
Premature leaf fall	*				*	*
Root growth reductions		*				
Plant growth reduction	*	*				*

	Nitrogen	Phosphorus	Potassium	Magnesium	Calcium	Sulphur
Deficiency most likely to occur	Intensively cropped land and low organic matter	Heavy clays, light sands and acid soils	Intensively cropped land	Lime-rich soils and low organic matter. Associated with heavy dressing	Acid soils	Only now showing deficiency
Affect crop yield	*	*	*	*	*	*
Bud dormancy prolonged	*	*				
Cure for deficiency	Nitrogen fertilisers, solid or liquid feed, quick acting	Phosphate fertilisers in root zones FSP seed-bed	Potash fertilisers, e.g. sulphate of potash, quick acting	In soil magnesian lime or Kieserite plant foliar spray of magnesium sulphate (Epsom salts)	Soil lime but check pH. Apply calcium nitrate spray at 14-day intervals	Use of sulphate fertilisers, e.g. super phosphate potassium, sulphate, etc.
Susceptible crops	All, except possibly legumes	Carrot, lettuce, broad bean, sweetcorn	Spinach, leek, lettuce, cauliflower, radish	Tomato, cauliflower, cabbage, marrow	Celery, lettuce, brassicas	Cabbage, leek, swede

TABLE 3   TRACE ELEMENT DEFICIENCY SYSTEMS

Nutrient	Occurrence	Symptoms	Remedial Action	Susceptible Crops
Manganese	Peaty soils above pH 6.5 and calcerous soils with poor drainage. Tractor wheelings	Interveinal chlorosis on new leaves	Avoid over-liming; foliar spray of manganese sulphate	Pea, french bean, onion
Iron	High pH and water-logged soils	Interveinal chlorosis. Young leaves may be bleached white	Iron chelates	Fruit – especially apples – sweetcorn, brassicas
Copper	Peat soils, especially recently reclaimed peats and organic chalky soils	New leaves greyish green, chlorotic and wilting	Copper sulphate applied to soil	Carrot, lettuce, onion, brassicas, celery
Boron	Light alkaline soils in high rainfall areas	Cracking of stems and petioles, distension of leaves, 'hollow stem' in brassicas	Borax to soil or solubor foliar spray	Brassicas, celery, beet, carrot, lettuce
Molybdenum	Rare except on acid soils below pH 5.5	'Whiptail' of cauli-flowers, i.e. new leaves only have midrib. Plants blind	Raise pH to 6.5 and/or sodium or ammonium molybdate	Cauliflower and other brassicas
Zinc	Not recorded in UK			

Boron (Bo)

Boron deficiency is most likely to occur on alkaline soils especially in dry seasons. Cauliflowers and Chinese cabbage seem particularly sensitive. The problem can be rectified by applications of borax.

Molybdenum (Mo)

This shows itself as 'whip tail' in brassicas, particularly cauliflowers. It is most prevalent on acid soils particularly below a pH of 6. Liming will normally correct the deficiency by raising the pH, or sodium molybdate can be used.

Zinc (Zn)

A zinc deficiency is very rare and only associated with a very high pH. It can be applied inadvertently to soils, as can other heavy metals, via sewage sludge, to a toxic level. This has actually occurred and fields so affected have become unusable as it cannot be removed from the soil. So if using sewage sludge please check for heavy metal contamination before spreading on the land.

Table 3 lists the nutrients, where the deficiency is most likely to occur, common symptoms, and recommended remedial action.

FARMYARD MANURE

I have already talked much about the soil-improving qualities of farmyard manure (FYM). Being organic, it has an important role to play in nutrition as well. Its main constituents are straw and dung, and as these are of vegetable origin they will return a complete range of nutrients to the soil. There is one possible problem with any form of organic matter: its nutrient content can vary considerably. It will vary with the ratio of straw to dung and whether it has been composted or kept. It even varies according to which animals produced the dung, e.g. chicken manure is very 'sharp' or hot, as it has a high proportion of ammonia. Thus, it is difficult to ascertain at what rates the various nutrients are being applied. It is common practice, though, to reduce fertiliser dressings if the land has had a reasonable dressing of FYM.

With crops where an accurate balance of plant foods is necessary, it would be better to use only well-composted or lower levels of FYM to minimise the variation in nutrition levels. Having said that, the benefits of regular dressings of FYM far outweigh the disadvantages.

5 Cropping Possibilities

Having covered crop requirements in the last chapter we now need to examine the crops themselves. Before plans are made to grow specific crops, they should be considered in the overall context of the marketing plan (*see* Chapter 2). If the grower has spent time formulating a method of marketing this in turn will suggest broad areas of cropping possibilities. For example, if growing for a supermarket outlet there is no point in diversifying into a crop which may totally change the marketing system. Growing crops and marketing are so interlinked that they cannot realistically be separated. Someone growing for their own retail outlet or farm shop will, of necessity, grow a much wider range of crops than might otherwise prove economic. If running a Pick Your Own enterprise only crops suitable for such marketing methods make sense.

In this section I will consider each group of crops and look at their general requirements and marketing possibilities. For detailed information about individual crops ADAS and its publications, and the major seed houses can provide an endless source, as can the *Grower Guides* published by *The Grower* magazine.

VEGETABLES

This group covers a very wide range of individual crops. To make life easier this range of crops can be grouped into three main classifications: brassicas, legumes and roots. This classification has many advantages. It is useful from the nutritional aspect, as crops within these groups will tend to have similar requirements. It is also useful from a pest and disease aspect as it allows rotation of cropped areas which prevents a build-up of soil-borne problems. This classification can also aid the management of the holding.

Brassicas

This group includes all types of cabbages, cauliflowers, sprouts, broccoli, swedes, turnips, radish and kohl rabi. All these crops prefer an alkaline soil (it helps to prevent club root), and to produce quality crops they need to be planted into firm soil. These two requirements combined aid the management of the holding, enabling the grower to produce the conditions required by the crops on a larger area of land; hence production costs are reduced by efficient use of machinery, chemicals and fertilisers, etc.

It makes sense, if you are liming in order to maintain the correct pH (*see* Chapter 4, page 42), to follow this operation by planting the crops which will gain maximum benefit from it, i.e. brassicas.

I have already mentioned that vegetables of this group like a firm soil – in fact, very few plants like their roots in a settling soil – but growing most brassicas in a light fluffy soil will produce light fluffy sprouts and loose-headed cabbages. In the old times they used to reckon a brassica was firmed enough at planting when it could not be pulled out of the ground with a sharp tug – the leaf ripped instead! Freshly incorporated manure in the ground to be used for brassicas is also to be avoided as it can produce soft heads. This is due to an imbalance of nitrogen and potassium. Though the cabbage family is generally considered to be a gross feeder, the correct balance of nutrients must be maintained to produce good quality of crops. Potassium is especially important for cauliflowers.

The leafy brassicas and cauliflowers especially are very susceptible to any check in the growing regime, particularly after planting out. With care, the plant can be controlled if a modern plug seedling is used but, after transplanting, any check can reduce the quality of the product.

Well-established, young brassica crop.

Cropping Possibilities

It is possible to programme harvesting of brassicas over a prolonged season with the aid of the many excellent modern varieties. Most brassicas that have been bred for the commercial grower are F1 hybrids. These have advantages in evenness of cropping, i.e. all from a similar sowing/transplanting date will mature over a few days. This can greatly aid mechanisation and harvesting operations, and in providing a continuity in volume if your market requires that. But it could be a disadvantage in a farm shop situation, as you may have a glut for a week and then no more. In this latter situation the use of 'open pollinated' varieties which are more variable in harvest date may be more appropriate.

Spacing of the plants is another very important aspect: by varying the spacing a particular size of cabbage can be produced. So check what size of head your market requires and match this to a variety and spacing that will produce this size. Many modern varieties can actually be grown closer than is at first thought normal due to advances in plant breeding, and the market (in general) appears to prefer a smaller product. By varying the spacing it is possible to grow to your customers' requirements, be it retail, catering or supermarkets.

To summarise then, brassicas prefer a fertile, well-drained soil that is firm, and they benefit from additions of lime. Cauliflowers grow best on lighter land than cabbages and sprouts.

It is impossible to give accurate recommendations for fertiliser applications. For a general guide, a phosphate and potash application of 50 to 100 units is required, but cauliflowers may need 150 units of potash. Nitrogen requirements, as we saw in the previous chapter, depend on previous cropping but 150 to 200 units may be needed; however, don't give any brassicas too much as it will cause problems. It is best to split this dressing into two applications.

Herbicides can be used for weed control as can mechanical methods. Weeds are not usually a problem once the foliage touches from plant to plant, so the use of a stale seed-bed will be very useful as this will help weed control and ensure firm soil.

Brassicas as a group offer a good marketing potential, for it is possible to supply some form of them over a very long period of time. As large acreages of brassicas are grown in Lincolnshire they are probably not best suited for a market-garden crop except to supply a local market or retail outlet where the freshness of a local supply can be advantageous.

Pests and Diseases I have already mentioned club root as a problem. Good general hygiene, including sterilisation of propagation areas and equipment, and care in the use of bought-in plants on the holding can all play an important part in ensuring this disease does not infect your land.

Organic pest control on brassicas. Tagetes has been planted to dissuade aphids, etc.

It can be brought in on machinery and even on boots so it may only be a matter of time.

Other diseases which can cause problems on brassicas include Alterneria leaf spot. This usually appears from July to September and initial infection seems to coincide with harvesting of oil-seed rape in which Alternaria can be endemic. At present, Rovral is proving an effective chemical for control. The diseases ring spot and leaf spot (not to be confused with Alterneria) can be problematic in wet weather from August onwards. Using a wider spacing may help to control these two diseases, as will Benlate. Pseudomonas, a bacteria which causes black water-soaked lesions can be seed-borne. It is spread by water splash and can get into a crop from an infected water supply. Good hygiene and efficient removal and burial of trash assist a great deal in preventing major problems as pools of infection cannot build up.

Pests of brassicas include the dreaded cabbage-root fly. This pest is attracted by smell to the brassica plant, be it crop or weed, and lays its eggs at the base. The hatching grubs then eat their way into the plant causing damage and death. There are in fact three generations of cabbage-root fly. The first coincides with cow-parsley flowers – usually for a five-week period in May-June. Only young plants are susceptible, and modifications to the planting programme can help to overcome this problem. The second generation emerges in July–August; and the third, which does not always occur, comes in September and is normally only a problem on sprouts.

Other pests include aphids and caterpillars. These can be controlled either by the use of chemicals or by organic means. Flea beetles can be a problem too with field-sown crops.

Legumes

For the purposes of this classification the group legume includes onions, leeks and spinach as well as peas and beans. This is so because some of the general requirements of onions and leeks are similar to the genuine legumes. It is also common practice to group these crops for rotational purposes depending on the areas of each individual crop being grown. Lettuce and celery can also be included in this grouping if being grown outdoors but I will deal with them in detail in the Salad Crops section (*see* page 63).

Leguminous crops are generally known for their ability to 'fix' free nitrogen in the soil due to the bacteria contained in the root nodules. This could indicate that they do not need either nitrogen fertilisers or the use of organic matter. In reality just the opposite is true.

All the crops in this group will benefit from dressings of farmyard manure of around 28 tons per acre (70 tonnes/ha). The manure should be well rotted, particularly for onions, as lumps of coarse farmyard manure can hinder root development in this crop. The organic matter will be beneficial in a number of ways. It will, as we have seen in Chapter 3, improve the soil structure, and all these crops like a deep, fertile, well-drained soil. Organic matter will also help retain moisture since all legumes benefit from a moist soil. It will also go some way to providing the nitrogen requirement of the crop. The ability of peas and beans to 'fix' this nitrogen into the soil and therefore make it available for a following crop can be used to advantage in the rotation by following well-manured crops such as these with brassicas.

The soil for this group needs to be cultivated in autumn and the organic matter incorporated at this time. The pH of the soil should be between 6.3 and 7.5. If lime is needed (ideally, it should be included in the rotation before brassicas), it can be applied after ploughing. Subsequent cultivations should be aimed at producing a deeply worked soil with no compaction. This latter point is particularly relevant to onions and leeks as they do not have a branched rooting system; a deep root system with any compaction will restrict the root run and give a poor yield.

Peas

Peas of the right variety can be sown in autumn (October) and will normally produce a slightly earlier crop, with main sowings being made in March through till May for succession. The selection of variety will depend on the site and the market the crop is destined for as well as the intended harvest date. Sugar-pea or mange-tout may be a good crop for the market garden as it is not grown widely and commands a good price.

Beans

There are many types of bean which can be grown and again I would advise growing those which best suit the market outlet. Broad beans are eminently suitable for Pick Your Own operations as in fact are peas. If growing for the wholesale markets, runner beans can be profitable if they can be trained economically; there is usually a good demand for quality runner beans even from the supermarkets. If growing for a retail outlet, pinched beans – that is, pinching out the tips of runner beans as they begin to 'run' – will produce an earlier crop and hence command a high price; but having said that, the quality of pinched beans is often inferior to a trained crop. For the earliest crops, raising the plants under

Autumn-sown broad beans in early March.

glass and transplanting them when frosts are over may give an early advantage.

French beans need to be sown when the soil is sufficiently warm. Drilling too early can result in very disappointing germination and growth. In southern areas early May is best, though by using some form of protection this could be brought forward. Climbing French beans can be a more specialist crop and can perform well under cold glass.

Weed control for peas and beans can be done both mechanically and chemically, and of course the land should be free from perennial weeds.

Pests and Diseases   These include the dreaded black bean aphid. Control measures for this common pest include removal of the soft growing shoots after the first pods have set. Late or delayed crops are the most susceptible. If using chemical control, care should be taken as bees polinate the crop. If 5 per cent of plants in the south-west corner are infected, that is the time to instigate the spray programme. Chocolate spot can be a serious problem on broad beans. To reduce the risks, ensure that the site is well drained and neither too well sheltered nor too exposed. Autumn-sown crops are most susceptible, particularly in humid conditions. As with all crops, good hygiene must be practised. Halo blight and anthracnose can both be seed-borne. The latter problem can be serious and if it occurs no beans should be grown on that site for several years.

Onions and Leeks

Accurate drilling of all onion crops is essential for an even stand, and the seed-bed should be prepared with this in mind to give a fine tilth. Spacing of rows will depend on the means of weed control to be used. Chemical weed control will allow closer spacings. No chemical herbicides should be applied between crop emergence and past crook stage.

This problem of weed control is also true for leeks, i.e. no chemicals should be applied from germination to past crook stage. For this reason, leeks are often sown in a stale seed-bed and then transplanted. This use of transplants makes weed control and crop management much easier. Both onions and leeks like a soil with a pH of 6.5 to 7.0 and the use of 30 tons per acre (74 tonnes/ha) of farmyard manure (well-rotted) will be beneficial. Spacings, especially the widths between rows, will depend on the means of weed control being used.

Onions are normally lifted in late August and allowed to dry in the field (weather permitting), and then sold or stored for sale from September through till March. To keep the crop later than March specialist storage areas are needed.

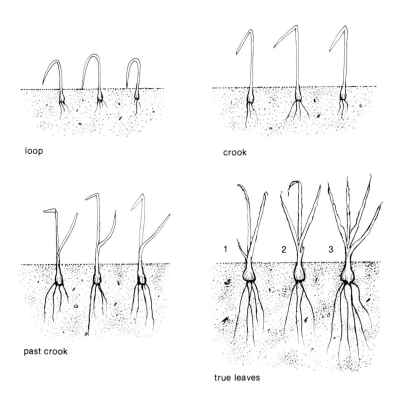

loop

crook

past crook

true leaves

Fig 8   Stages of development of onion/leek seedlings.

Leeks can be produced for harvest from late August through to early May by the selection of suitable varieties. Timing of sowing will also affect harvesting date.

Pests and Diseases   Pests are not normally a problem on onions and leeks but eelworm and onion fly can sometimes be troublesome. Good hygiene and a rotation of a minimum of three years will help. For the fungal diseases such as white rot and downy mildew an even longer rotation should be practised (five years' minimum) and all infected plants and crop debris burnt. For details of chemicals to use it is best to contact the local adviser. Rust occurs on most leek crops; the most common source of infection is crop debris and volunteer plants in other parts of the holding. Leek moth is also starting to be a problem in certain areas.

Cropping Possibilities

A good crop of leeks, well ridged up.

Spinach

This is another crop that can be included in this group. The benefits of having a well-manured soil for spinach include the available nitrogen and the added moisture retention as spinach is susceptible to drought. Spinach is a more specialist crop and should be grown only for a specific outlet. It is possible to crop from April till November from outdoor planting, and October till April under glass. Soil pH needs to be 6.5 to 7.0 and spinach should be grown on well-drained, moisture-retentive loams.

Roots

In this section we will consider carrots, parsnips, beet and potatoes though other specialist or unusual crops such as salsify and scorzenera would also come into this section.

Roots in general prefer a light sandy soil which is stone-free. Compaction can cause bent and forked roots, and for this reason a bed system is often used. Organic matter should not be incorporated for root crops as it will induce forking. In drier areas irrigation will be beneficial, though water should be applied consistently or the roots will split.

The basic nutritional requirements of this group of crops is for a low level of nitrogen but a relatively high level of phosphate and a medium level of potash. If we look briefly at this requirement in association with that of brassicas and legumes, we can see how the usual rotation of legumes, brassicas and roots has developed.

1st Year
Farmyard manure and Legumes

2nd Year
Lime and Brassicas

3rd Year
Fertilisers and Roots

4th Year
Farmyard manure and Legumes

Carrots and Parsnips

Carrots grow best in sandy stone-free soils or the Fen peats. The pH should be in the range of 5.8 to 6.8. The crop can be produced for sale at any time of the year, though to have roots for sale in June requires controlled storage and is probably beyond the means of a market garden. It should, however, be possible to produce fresh roots from under glass if space is available. Roots for sale over the winter period (November–April/May) can either be stored under cover or *in situ* in the field. For the latter, a good cover of straw – 15 inches (38 cm) deep – is needed and for late storage the straw should be covered with black polythene.

As has been stated, all efforts should be made to avoid soil compaction. A loose seed-bed produced by a once-over pass is best. Fertilisers are best applied at ploughing though nitrogen should be applied as a top dressing when the crop can utilise it. The crop is normally precision drilled to a given spacing. The exact spacing depends on the size of root required by the market outlet, as plant density has a major effect on the diameter of the root. Variety will depend on sowing date and shape required. If you are not sure what variety to grow, the National Institute of Agricultural Botany (NIAB) produces a list of many vegetable varieties and their uses which may be a useful alternative source of information from seed catalogues. The crop responds well to the use of floating films (*see* Chapter 7, page 101), but if using such techniques to gain early production, ensure it is removed at the 6–7 true leaf stage. If using film, it is usual to sow the drills in hollows to keep the seedlings away from condensation on the film.

The above details are similar for parsnips and carrots. Both these crops are normally marketed in 28 lb (12.7 kg) bags or nets. There may

Cropping Possibilities

Carrots under extensive floating film, on coastal sands in Suffolk.

be marketing opportunities with catering establishments for baby carrots and even baby carrots with foliage over a larger season.

Pests and Diseases   The first one that springs to mind is, of course, carrot-root fly. This pest is very like cabbage-root fly. It produces three generations in most seasons, the first at the end of May, the second in mid-August and the third at the end of September. In high-risk areas the use of chemicals will not give complete control. The real key to preventing this problem is isolation. If you are lucky enough to be at least two miles from any other carrots, parsnips, celery or parsley, you

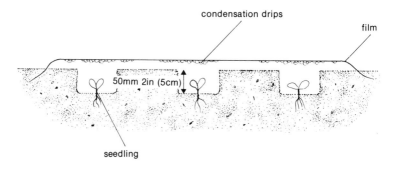

Fig 9   Recessed drills under floating film.

may not have a major problem. Sheltered fields are also more susceptible to carrot-root fly damage. In fact, the first generation can be missed by timing sowings, so that the crop emerges after the risk of infection is past. It is the second generation that causes the real problems. Attempts at chemical control can be made by the use of granules at drilling and sprays in mid-season. The variety Nardor is reputed to have some resistance to carrot-root fly. For the organic grower covering the crop with a material such as Agrinet at high-risk times may help, and smell-hiding compounds are said to be useful as well. The third generation is only likely to cause problems in crops stored in the field after early October.

The fungal disease of cavity spot can cause problems but can be controlled chemically. Violet root rot is a much more serious soil-borne fungal disease, and if this is found on your land carrots should not be grown for ten to fifteen years. Mayweed and chickweed are also susceptible to violet root rot.

With parsnips the main problems are canker and carrot-root fly (*see* above). The variety Avon Reister is resistant to all forms of canker and should be grown where the disease is a problem. Other cultural methods which may help include good drainage, closer spacings, later sowings and earthing up in summer. The use of clean seed is also recommended.

Beetroot

Beetroot is only really worth growing for an established fresh market outlet as the returns on this crop can be low. It is normally sown from April to July, though there is a tendency for the early sowings to bolt (run to seed). This is due to a combination of low temperatures and long days. Some varieties, such as Boltardy, have resistance to this problem. When drilling, if using a precision drill, the use of rubbed seed will improve the number of singles sown. If mechanical weed control is to be used, the incorporation of some radish seed will enable the rows to be seen as beetroot is slow to germinate. Spacing, both within the row, and of rows, will depend on the size of roots required by the market outlet. Soil conditions should be as for the other roots. Irrigation is necessary for top quality as beetroot is a shallow-rooting crop and wilts easily in dry weather. The normal harvest period is from July to November and the crop can be stored for winter sales using a clamp made of straw bales.

Cropping Possibilities

Potatoes

The other main root crop is the potato. On many intensive market gardens there may not be room to produce potatoes economically, but there may be marketing opportunities which cannot be ignored. Specialising in salad potatoes or unusual varieties such as Pink Fir Apple may be worth consideration. Potato growing and marketing is controlled by the Potato Marketing Board but this should not be a problem when small areas of the crop are grown. Potatoes are often used as 'ground breakers' as they like an organic-matter enriched soil and are often used as a first crop in a rotation.

Always use certified seed; never use twice-kept seed. The seed should be chitted early (= left in a cool, dry, light place to develop sprouts), and hand or machine planting can be used. Potatoes prefer a soil which has had minimal cultivation, with a pH between 5.5 and 6.5. The crop is responsive to phosphates, and potash levels should be kept low. Earlies should be planted in early March when the soil temperature is above 39°F (4°C).

Weed control is only necessary until the crop covers the ground, and it can be done either chemically or mechanically by ridging.

Potatoes are very responsive to irrigation. A minimum amount of water of 1 inch (2.5cm) per week is required, either from rain or irrigation, from the tuber marbling stage. Harvesting can be by hand or by machine but potatoes should never be dropped more than 6 inches (15cm).

Pests and Diseases These include blight, scab, wire-worm and nematodes. Blight can infect any crop after a 'blight period' of warm damp days. This disease can only survive on living plant material so hygiene can help. If it does occur do not irrigate and if tubers are sufficiently large destroy the haulm to prevent spread. Chemical sprays only delay the inevitable but may take the crop past harvest. Scab can be a problem, particularly in alkaline or limed soils but the use of irrigation will help to control it. Wire-worms are only usually a problem if the crop is following permanent pasture. Nematodes or eelworms can be a problem and the use of a good long rotation will help prevent the build-up of this pest.

There are some other vegetables that may be considered suitable market garden crops. These include the marrow family, sweetcorn, and more unusual vegetables such as kohlrabi, fennel, scorzenera and celeriac.

Marrows, courgettes and the less common squashes and pumpkins all

Sweetcorn, an ideal Pick Your Own crop.

like a soil rich in organic matter, and the ability to irrigate the crop can be an advantage. When considering these crops the areas grown should be evaluated carefully for some crops may take up large areas (pumpkins, etc.), while others have a very high labour input (especially courgettes, which really need harvesting on a daily basis to prevent the fruit becoming too large).

Sweetcorn lends itself to the Pick Your Own enterprise. This product suits the smaller grower and can give good returns.

SALAD CROPS

This group of crops could be very important to the market garden. Once again, crops which are grown on an extensive scale such as lettuce and celery should perhaps be grown only if for a specific outlet, as market prices and volumes question the economics of these crops. More unusual varieties of lettuce such as Little Gem and the endive and red varieties may be suitable crops. Round or butter-head lettuce can be produced outdoors from May to October.

Plants are normally raised in peat blocks and transplanted. Since the

Lettuce seedlings in peat blocks.

growing period is relatively short, a continuity can be achieved on a relatively small area. Crisp or Iceberg lettuce takes a little longer and cropping of cos types is more restricted. Lettuce is often used as a 'bread and butter' crop on many modern market gardens but be prepared to plough in some crops if market prices make harvesting uneconomic. Lettuce are also susceptible to a wide range of pests and diseases, but chemical controls are readily available.

Celery, as mentioned above, is possibly not a crop for the market garden but can be grown successfully outdoors using the self-blanching types. Irrigation is a necessity for this crop, as is a good, rich, moisture-retentive soil.

Radish and salad onions can easily be grown on small areas and in continuity. I know of some market gardens that have specialised in these crops and have been very successful in supplying national super-market chains.

For early production of all these crops, light land with a favourable southerly aspect is best and irrigation a must to reach the quality standards now being demanded.

Protected Cropping

There is quite a range of crops that can be grown for most outlets under protection. For these purposes glass is best but good crops can be

*Protected cropping: tomatoes and aubergines in small
poly-tunnel.*

grown under polythene. The crops that can be grown in this way
include tomatoes, cucumbers, peppers, aubergines, melons, lettuces,
radishes and celery. These crops can be divided for ease into trained and
untrained (or ground) crops.

When growing under protective structures of glass or polythene
great attention must be given to the soil. Most protected structures are
cropped very intensively and this can cause problems if care is not taken.
The soil structure needs to be improved to maintain a good root zone
so the regular addition of organic matter will be beneficial, as will
regular liming. Under protected structures regular liming will be
necessary as the use of fertilisers, liquid feeds and farmyard manure will
all acidify the soil, even in hard-water areas. The pH should therefore be
checked regularly and adjusted as necessary. Regular subsoiling and the
use of deep cultivations occasionally will also be useful in extending the
root zone.

One result of this intensive cropping may be a build-up of soil-borne
pests and disease, especially as a rotation is often impractical under
protection. This could mean that the soil will need some form of
sterilisation at regular intervals which can be done either by chemicals
or by steam sterilisation. It is now possible to buy or perhaps hire small

Cropping Possibilities

Irrigation in poly-tunnel. Note end door for ventilation.

mobile steam generators which should make this viable, certainly for small areas.

Two other factors need to be mentioned since they play a vital role in protected cropping: ventilation and irrigation. Ideally, the ventilation area in a glasshouse should be between one third and one sixth of the floor area. It is in fact often smaller and this can cause problems during the season. Irrigation systems also need to be good, and well maintained. A crop of summer lettuce under glass will soon show up any weaknesses in an overhead irrigation system. It can also be of great advantage to be able to liquid feed via both an overhead irrigation system and a low-level or trickle system. Finally, crop supporting wires need to be substantial and well maintained, and securely attached to strong ends of the structure. I have heard of a glasshouse that had grown tomatoes successfully for many years but when a crop of cucumbers was grown the ends of the glasshouse were pulled in! One tip – keep crop support wires in as short a section as is practical for two reasons: firstly, they are not supporting such a weight, and secondly, if they do break they are easier to repair. Mine was 150 feet (45 m) long when it snapped!

Quick coupling to allow liquid feeding via overhead spraylines.

Cropping Possibilities

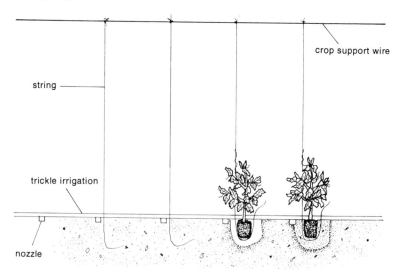

Fig 10 Stringing and planting tomatoes. The string is placed under the root ball.

Tomatoes

Tomatoes are probably the single largest crop grown under protection but from a market-garden point of view they will probably only be grown for sale at a local retail point. There are several types that can be grown these days: the normal round types, beef or marmande types, cherry tomatoes and even plum and yellow tomatoes.

The training system is similar for all types. Unheated crops are usually planted in April when the glass will keep out any late frosts. The plants should have the first few flowers open at planting. I liked to prepare the house fully before planting as this made later care easier. After soil preparations, the trickle irrigation was laid. The individual strings were then tied to the wire, the length being about 18 to 24 inches (46–61 cm) longer than that needed to touch the floor.

When we planted, the string was passed under the root-ball. The tension was sufficient to hold the string, especially as I gave the first twist at planting as well. I found this procedure meant that I did not have to return to the house for training for possibly two weeks until the plants had grown sufficiently to warrant twisting around the string and side-shooting. I also removed as many side-shoots as possible at planting.

All watering was done by trickle irrigation and feeding started after about three weeks, depending on crop growth. The crop should be sprayed over with a high-pressure water jet between 11 a.m. and 1 p.m. to aid setting of early fruit. Some growers use an 'electric bee' for this. The crop should be deleafed to aid air circulation and enable fruit to be seen for picking. If growing cherry tomatoes, leave approximately 3 to 4 feet (90–120 cm) of foliage on the plant as you will be picking from the first and eighth truss at the same time!

On reaching the support wires, the plants can either be stopped, i.e. the tip pinched out, or trained along the top wire or across the path, etc. Stopping should take place six weeks before it is intended to remove the crop to enable all the fruit to ripen.

When growing cherry tomatoes, such as Gardener's Delight, certain modifications are needed to a normal feed regime. To gain maximum flavour, use about half the volume of water of a normal crop and always use a high potash liquid feed. Care is needed to balance the water and feed requirements in order to control the crop. Too much water and it will go wild, too little and it will stop growing altogether.

Pests and Diseases   The main pests and diseases for the crop are whitefly, red spider mite and botrytis. I have managed to control both the pests successfully by biological means. They can of course be controlled by chemical means, resistance can be built up so use at least three insecticides. Alternating both systems can help. Red spider mite will prefer hot dry conditions so damping the crop over can help reduce the numbers. Having said that, a damp atmosphere is ripe for an infestation of botrytis. This can be limited to a greater or lesser degree by cultural methods. Regular deleafing, general hygiene and good ventilation all help to reduce the incidence, as will ensuring that the crop is dry at the end of the day. It is important to control botrytis as even at low levels of infection 'ghost spotting' can occur on the fruit and detract from their quality. Other diseases can be a problem, especially some soil-borne ones such as fusarium. If they become a major problem it may not be possible to cure them by sterilisation, and the crop will have to be grown in grow bags instead.

Cucumbers

Traditionally, cucumbers were grown on straw bales. This practice is not necessary for short-term crops but a soil rich in organic matter will be of real benefit. Cucumbers are gross feeders and require large amounts of nitrogen. They come into cropping in a relatively short period and can be quite prolific. A good crop can be grown in the older types of glasshouses, but remember the crop can be heavy and support wires need to be good. Plant them a little later than tomatoes when the soil is warmer. If trying to produce over a long period, replanting in mid-season or interplanting could ensure continuity of supplies. The use of all female flowering varieties will eliminate the problem of bitter fruit. For maximum weight of crop do not let any fruit develop on the bottom two feet (60 cm) of the main stem. The plants can be trained either as cordons with the fruit on the main stem or on short laterals, or by the umbrella method, i.e. a single main stem with several branches at the top.

Pests and Diseases   Pests and diseases are very similar to tomatoes, but extra care is required when using chemicals as cucumbers are susceptible to some. Whitefly is probably the biggest problem as a pest, and mildew as a disease. The mildew is a powdery type which prefers a dry atmosphere, so increasing the humidity will help; however, this encourages botrytis if overdone.

Peppers and Aubergines

I have grouped these two crops together as their basic requirements are very similar. They should not be planted out too early as they like to go into a warm soil and our experience shows there is little to be gained from planting before the end of April when no heat is available for the crop. Some form of training is required to support the plants. It can range from a single cane or single string, like tomatoes, to four strings to support each branch. One advantage of these crops is that they do not have such a large labour requirement as tomatoes or cucumbers since training and picking only needs to be done every ten to fourteen days. Feed requirements are basically a warm rich soil and liquid feed, with a medium nitrogen feed at each watering. Blossom drop can be caused by uneven watering and the crops benefit from damping down. The first 'king fruit' on peppers may need to be removed as it can often become wedged between the branches.

Pests and Diseases   Red spider mite can be a problem, particularly with aubergines. Caterpillars can be damaging in a crop of peppers and can be difficult to spray as the foliage is so dense.

Lettuce

Lettuce can be grown all the year round under glass though careful selection of varieties is required. Spacing varies, but is most commonly 9 × 9 inches (23 × 23 cm). Lettuce requires a good rich soil with a pH of

Lettuce seedlings which will be ready for planting at the 3-5 true leaf stage.

6.5. Soil conductivity should be low and, if lettuce is following a crop such as tomatoes, flooding may be necessary to reduce the levels of salts in the soil.

Hygiene is very important with continuous production, and all crop debris should be removed from the glasshouse. Regular sterilisation is needed. Ventilation of the crop is critical, and air movement can help to reduce the problems of glassiness and will help prevent rots, etc. Good overhead irrigation is required; if there is any unevenness in pattern of the nozzles this will soon become apparent, and an uneven crop will result.

Paths should be kept to a minimum but all parts of the crop should be accessible for spraying with crop protection chemicals. On my own holding I used a motorised knapsack sprayer for all insecticides and fungicides which proved very effective on any crop.

Care should be taken when harvesting the crop, and all harvesting should finish by 10 a.m. in summer to prevent a build-up of field heat in the product. This applies to all leafy crops, as it is field heat that reduces the shelf-life of the produce.

Pests and Diseases Many pests and diseases can affect lettuce. Continuous production can greatly increase the risks of problems occurring. Good hygiene must always be practised; any rubbish heaps must

71

be located well away from production areas. The most common problems include aphids, botrytis, tip burn and a range of bottom rots. With fungal diseases, keeping some air movement round the crop, and watering only in the mornings will help reduce the instances. Mildew can also be a problem and resistant varieties should be grown where applicable. A range of chemicals should always be used to prevent a build-up of resistant strains.

Celery

Celery can be grown under glass for harvest during May and June. For the early crop, heat or some form of frost protection is needed. When propagating celery plants for early production ensure that they do not get checked, and are propagated at relatively high temperatures as this helps to prevent bolting. Once again, a good irrigation system is required as the crop has a high demand for water. The crop benefits from the incorporation of high levels of farmyard manure prior to planting although this can cause problems with slugs and weeds. Both can be controlled chemically, but if using herbicides under glass great care is required as this practice is not normally to be recommended.

The crop will require feeding with nitrogen, either by using a nitrogen liquid feed at every watering via the overhead irrigation, ensuring it is washed off the foliage, or by hand. If using the latter system, I find that a watering can with a short length of hose on the end of the spout makes an ideal applicator. Two solid dressings should be applied, one when the crop is established and one at the last possible time the crop can be walked over. The fertiliser only needs to be applied to every other row, but take care it does not touch the plants. With self-blanching crops a spacing of 9 × 9 inches (23 cm) will be suitable.

The main labour input on celery is at harvest as all the crop will need to be cut in a relatively short time for top quality.

Pests and Diseases The crop is relatively free of pests and diseases but aphids can be a problem, especially round any weeds that appear. Slugs can be controlled by adding slug pellets to solid top dressings or at planting time. Sclerotinia disease can be a problem; if it occurs, infected plants should be removed carefully and destroyed, and the soil sterilised after harvest. Once on a holding, it cannot easily be eradicated but can be kept to acceptable levels. The other major problem is leaf spot. This usually comes in on infected seed or plants, or via the water supply. Regular spraying can control it but initially removal of infected plants is to be advised. Outdoors and even under glass it can be spread by rabbits running along the rows!

Radish and Salad Onions

These can also be grown under glass both all year round and to extend the season early and late. Other possible crops include calabrese, endive and kohlrabi. These may make alternatives to winter lettuce.

HERBS

Herb growing has now a new lease of life with the change in eating habits and the rise in alternative medicine. Many people have seen herb growing as a new area for production. Careful appraisal of the marketing opportunities is required though before taking the plunge. Herbs can be marketed as a fresh product or as dried herbs.

For production, a good rich soil free of perennial weeds is required. With all herb crops, weed control needs to be carefuly considered as it may not be possible to use chemicals. This may be so because the market demands an organic product or because there is a lack of approved chemicals for use on the range of herbs being grown.

The crops will be labour intensive to harvest and need to be cut at the correct stage to maximise returns. Any drying must be done carefully to retain the aromatic oils, and may require specialist drying areas.

Having said all this, it may be that a market garden specialising in herb production could be very profitable, so it may well be an option worthy of consideration.

FRUIT

Extensive growing of fruit may not be in the field of market gardening but traditionally market gardens did grow some fruit.

Soft fruit (strawberries, raspberries, currants, etc.) lend themselves well to the Pick Your Own enterprise and make a welcome additional line to the farm shop outlet. If you are planning a Pick Your Own enterprise, consider the potential market and the number of people visiting as this will relate to the possible areas of each crop.

Strawberries are the most popular crop. Only clean certified plants should be used and beds will need replanting about every five years. There can be problems with strawing the crop and trying to keep runners under control, though I have heard of one Pick Your Own grower that used the 'matted bed' system.

Raspberries, blackberries and hybrid berries all need suitable supporting systems and regular maintenance. Currants and gooseberries

Early rhubarb.

Strawberries grown through plastic mulch.

74

Nursery stock (shrubs, etc.) being grown and sold at a farm shop/market garden.

are not as popular so relatively small areas of these should be grown at first.

Melons are not widely grown commercially as yields can be low, questioning the economics of the crop. Where they have been grown for farm shop sales the response from the public has been excellent and they may well be a crop worth looking at. Culturally they are similar to cucumbers and will need some form of protection, i.e. a glasshouse, tunnel or cold frame. They will require a higher level of potash to ensure flavour in the fruit, and more time in training as the fruit themselves may need supporting.

There may be a good opening for an organic Pick Your Own in many places.

Top fruit such as apples, pears and plums are really beyond the scope of a modern market garden but may again make a welcome addition to farm shop lines.

There are some other lines that can be sold via the holding's own retail outlet and can be produced with minimal specialist equipment. These include bedding plants, some shrubs and pot plants. However, all these are specialist areas and not really market garden crops.

6 Machinery and Equipment

In this chapter we will look at the range of machinery that is available to the modern market gardener. The actual equipment bought and used by the grower will depend on several factors. First and foremost of these will be money. Machinery is expensive, though by careful matching of machine to the job great savings in time and costs can be made. Other factors to be considered are the range of crops being grown and the soil type one has to deal with. There is not space in a book such as this to go into details of individual machines and makes, but by considering various options the grower should be able to make the best choice of equipment for his or her own situation.

PRIME MOVERS

This is perhaps an unusual heading but in reality quite logical for it is in this section that the tractor must be considered. The choice of these prime movers was in the past limited to the good old Fergie or Ford Dexta but in recent years the range of small and compact tractors has increased dramatically. The modern compact Japanese tractor makes an ideal tool for the market gardener. There is now also a wide range of implements available for these compact tractors, and their small size and manoeuvrability make them ideal for use under glass as well as in the field. The smaller standard tractors are also very versatile machines and are much more readily available second-hand, though they command a quite high price. When buying second-hand compact tractors, especially the early models, check the range of implements available as some of the early models only take specific equipment.

Another prime mover is in fact the rotavator. Several types can take quite a wide range of implements, including trailers. On my holding I used a Honda rotavator as my main power unit. By changing the rotors for wheels, it pulled a trailer and moved over 250 tons (254 tonnes) of farmyard manure and many thousands of boxes of produce round the holding over seven years. It was of course not legal for road use as it had no brakes (it was a flat holding!) but it did save the legs.

Prime mover I. A modern small tractor ideally suited to a market garden.

Prime mover II. Wheels can be substituted for the rotors, and a trailer attached.

A subsoiler, for pan busting.

CULTIVATION EQUIPMENT

There is a vast range of both types and makes of cultivation equipment and much of the general equipment can be bought readily on the second-hand market. Ploughs are the major primary cultivation tools of the farmer and grower. The size and type required will depend on the power of the tractor pulling it. It may be possible to use a contractor or a neighbouring farmer, etc., to plough the relatively small area needed. The timing of these operations can be critical and it may be advisable to do one's own work.

A subsoiler is a useful piece of equipment to have access to (*see* Chapter 3, page 26). Once again, match the size to the power available.

The range of secondary cultivators available includes rigid and spring tine types, disc and other harrows, all of which can be useful and should be matched to soil type and conditions. Used correctly on reasonable soil conditions they produce a workable tilth. The mode of action of tined cultivators produces a crumb structure which can help prevent capping or surface panning on susceptible soils. It may take several passes of the cultivator to obtain a good tilth and this can increase compaction. Powered cultivators, such as rotary harrows, can reduce the number of passes required but need a tractor with a relatively large power output to drive them. This may take them beyond the means of the market gardener.

A tractor-mounted rotavator.

The piece of equipment now most commonly used is the rotavator. Since its introduction it has revolutionised ground preparation: it chops up weeds and crop debris, breaks up soils and prepares a seed-bed all in one pass. It can be obtained in all sizes to match a tractor of any power output. It sounds a panacea for every ground preparation problem but this is not so. When a rotavator is used without due care and thought, real damage can be done to the soil structure. It is the one piece of cultivation equipment, if not the only one, which can create a

A pedestrian-operated rotavator at working depth.
This machine has been re-engined.

79

'cultivation pan' and destroy the vital crumb structure in the soil by its working. If used correctly, however, a rotavator can solve most cultivation problems. Care should be taken to match rotor speed with ground speed, and the soil type and conditions on the day. Remember, fastest ground speed plus slowest rotor speed to produce an acceptable tilth will not cause any soil damage. So exercise caution and only use one when the soil moisture content is at a suitable level.

Rotavators can be mounted on tractors either offset or parallel. An offset machine will remove the tractor wheelings on one side and hence wheelings can be removed from the field. The centrally mounted versions can be readily adapted as a bed-forming machine and are therefore best used when a bed system of production is being used. It is possible to fit 'pan buster' tines to follow the tractor wheels to reduce compaction. Some machines have been developed which have a series of small rotors and can be used for inter-row work, but these specialist machines will be expensive. To prevent a 'cultivation pan', the depth of use should be varied, or use tines on soils that are sensitive to panning.

WEED CONTROL MACHINES

Most weed control equipment relies on the hoe principle. Blades or tines of various types and designs are attached to a tool frame and are then pulled through the soil; like any hoe, they should cut the weeds off

A steerage hoe, for inter-row work.

A form of bed system. Rows left out for tractor wheels.

just below the soil surface. To make sure of this, the blades need to be checked and replaced regularly. These types of hoe can be used for inter-row work and, for this purpose, steerable hoes are less likely to cause crop damage.

Recently, the upsurge of interest in organic growing has fuelled development of other types of weed control as an alternative to spraying chemicals. Two developments that have come out of this are modern flame guns and brush weeders. Flame guns can be fuelled by paraffin or gas; brush weeders use contra-rotating brushes to scrub up the weeds. They actually do this without stirring up the soil too much and bringing more weed seeds to the surface, ready to germinate.

To use tractor-driven hoes or weeders, row spacings need to take the tractor's wheels into account. Bed systems can help reduce the potential for compaction and drainage from the tractor's wheels. Compact tractors are unable to have their wheels extended to give as good a bed width as tractors such as the MF 135 or Ford 4000. The ground clearance of the tractor will be important because the vehicle passes over the crop. For small areas nothing can beat the good old hand hoe. There have been several developments, though, the main one being some form of double-sided blade so that the hoe will cut on both the push and pull strokes. In between the two extremes, hoe and cultivator tines can be fitted to some pedestrian-controlled rotavators, such as the Howard and the Honda.

Machinery and Equipment

SPRAYERS

There is quite a range of sprayers suitable for use on the market garden, from small knapsack types to tractor-mounted types. A good sprayer is just as necessary as an organic holding for using organic materials such as 'Savona', soft soap, etc.

All sprayers should be regularly maintained to ensure they are applying the correct amount of chemical safely. Blocked or damaged nozzles can cause pollution and damage to crops by giving an uneven spread.

In an attempt to prevent crop damage it is a good idea to have one sprayer for crop protection chemicals, and one sprayer for herbicides. This prevents damage to crops due to incomplete washing out after herbicide use. In fact, on my own holding I used a standard CP3 hand-pumped 4-gallon (18-litre) knapsack for herbicide applications and a motorised air-blast knapsack sprayer for all crop protection work. I found this latter worked effectively both indoors and out, giving good cover of the crop, especially to the undersides of the leaves where most pests congregate. It was used with success on a wide range of crops, from lettuce to cucumbers and tomatoes to celery. The 'Turbair' system works well under glass though the range of specially formulated chemicals is limited. The next stage upwards from the knapsack is one of the barrow types of sprayer. These carry a larger tank and cover a wider area of ground. After that, there are tractor-mounted sprayers. These can be boom sprayers and may be adapted to enable a hand lance to be attached for spraying under glass, etc.

Great care should be exercised in calibrating your sprayer. It is essential to get the application rate correct, as too much can cause crop damage, is expensive and can pollute the environment. Too little on the other hand is ineffective and a waste of time and money.

The standard of spraying should improve thanks to the current FEPA legislation. This states that, as from 1 January 1989, all contractors and employees and all people born after 1 January 1964 have to have a certificate to spray the listed range of chemicals (*see* Chapter 8, page 107). This list includes all insecticides, fungicides, herbicides, and most organic preparations, growth regulators, etc. In fact, as a general rule get the certificate. The courses are made up of a foundation module which covers the law, health and safety, storage of chemicals, etc., and modules for each type of applicator. Chemicals are potentially dangerous so do have the training, and store and use them safely.

FERTILISER SPREADERS

I have yet to locate a suitable pedestrian-operated fertiliser spreader for use on a market garden. There are several on the market which can be either trailed behind or mounted on the compact tractors. Most are of the spinning-disc type or the oscillating-spout type. Once again, good calibration is needed to ensure the correct amount is applied, and this can be done either by spreading a known weight of fertiliser and measuring the area covered, or by collecting the fertiliser spread over a known area. The spreader will require calibrating for each material used. This sounds like a real chore but if done properly it will ensure economic use of fertilisers and, combined with regular maintenance of the spreader, will ensure that no problems occur with spreading or wastage.

Many market gardeners use the trusted 'bucket and chuck it' system. When applying fertilisers by hand it is best to split the dressing and apply the second part at right angles to the first to be certain of an even coverage of the ground. With a little practice it is amazing how quick and accurate application by hand can become.

SEED DRILLS

Precision drills (drills which sow a seed at a specific distance from its neighbour) are ideally suited to outdoor vegetable production. They are economical with seed and can be obtained as single hand-pushed units or in more sophisticated forms, up to multi-row, tractor-mounted versions. There is a range of seed-metering devices available, the most common being cell wheels and perforated belts. Both can be obtained in a range of spacings and for a range of seed sizes. When using precision drills it is wise to use graded seed. Cheaper, less accurate types of drill are available. To some extent the seed drill is in the process of being almost phased out as more and more vegetable plants are being raised and transplanted using the module system.

PLANTING MACHINES

Planting machines range from fully automatic machines capable of handling large numbers of plants per hour to some very basic devices which only mark out the ground.

Two of the basic types of machine may be suitable for use on a market garden. One is the tractor-mounted type of cabbage planter which has

Fig 11 Rotavator with marker drums.

been around for a long time and can be found second-hand. These machines usually have a system of fingers or discs to hold the plants, and they can be used and adapted for a wide range of plants. The modern module planters are probably out of reach financially for the normal small market garden but there may be second-hand machines available.

A third type of planting machine available is the lettuce planter. This consists of a pair of rollers on a common axle on which are placed bands with marker lumps at a regular spacing. Behind the rollers is a frame with seats for the operators. A rack at the front and back on which trays of plants can be stacked completes the machine. It can be driven either by a small petrol engine or by an electric motor, and there is a definite knack in steering them. With two skilled planters on a machine of this type, up to 20,000 plants per day can be inserted. The bands can be moved to allow different spacings to be achieved.

If these machines prove too expensive cheaper alternatives are available; I had some 'drums' made up to fit on the Honda rotavator at a very reasonable cost. This device marked out and, if the soil conditions were reasonable, made a hole in which to drop the block-grown plant. I have used this system – and a planting machine – for round lettuces, icebergs, celery, cabbages and leeks, and it could be used for any block – or even module-raised plants. A hand-pushed roller with lumps can also be used.

An elevator-type root lifter.

HARVESTING EQUIPMENT

Some very specialist harvesting equipment has been developed, but for the market gardener an elevator type of root lifter and/or a lifting blade will be more suitable. Perhaps the best, or certainly the cheapest, harvesting machine is the Pick Your Own customer! There are other machines which in reality are aids to harvesting. In this group I will include such equipment as the lettuce elevator which can be used on a wide range of crops and can lessen damage and ease the grading and packing operations.

GRADING AND PACKING EQUIPMENT

With certain crops some form of grader will be necessary to be sure of conforming to EEC standards. Tomatoes are an obvious crop. Once again, a small area of a particular crop may make the economics of such grading equipment a farce. One answer may be to join a local marketing co-operative. Other ideas have run on the lines of a mobile packing station which could be organised round an elevator. Another system is to have a large netted 'basket' at the front into which the cutters drop the produce. The packer then takes the produce out, weighs it if necessary and packs in the appropriate box. Pallet-wrapping devices may also be useful if produce is being moved on pallets.

A cold store is well worth considering. It will prolong the shelf-life of the produce harvested as it will remove the field heat as soon as possible, and will enable the produce to reach the customer in better condition than would otherwise be possible. There are three ways of procuring a cold store. They can be purchased as a unit, existing buildings can be converted or an old refrigerated lorry back or container can be used. With the latter, it needs to be adjusted to maintain a temperature of about 39 °F (4 °C). When considering the possibilities, a straight-through system should be adopted if possible. This enables freshly harvested produce to come in one side and cooled produce to leave the other. The use of a cold store will be a necessity if supplying a multiple chain or possibly a marketing organisation supplying similar customers.

TRANSPORT

This can be divided into two sections:

a)   On-holding transport. Some form of transport is needed to move things around on the holding. If a tractor is available a trailer and a transport box will be extremely useful. Another common means of transport is a tractor-mounted pallet lift. Standard pallets or pallet bins can be used to move a range of objects, and, when using them, some means of moving them round the packing shed or cold store will be needed. It may also be necessary to have some means of loading directly onto lorries. Some form of personal transport may save a lot of walking time. An old van, pick-up or estate car can be used but may be limited to dry days. I knew of one holding where the grower used a moped to get round his holding. A modern ATV three- or four-wheeled bike would be ideal but may not be fully justifiable.

b)   Off-holding transport. This form of transport must obviously be road legal. It may be possible to use the same vehicle for on- and off-holding transport. What really determines the best mode of transport is the use to which it will be put. If you have to transport produce to customers, a van of some sort may be most suitable; if delivery is to a local pack house just down the road, a tractor and trailer or pallet lift may be adequate; while if the produce is collected direct from the holding no transport may be needed.

IRRIGATION EQUIPMENT

Irrigation is vital in order to obtain the maximum yield from your land. With our recent spate of poor wet summers, irrigation may not be thought to be so important. In reality, most crops will benefit from irrigation during their lives – when sowing or transplanting, to ensure maximum growth, to aid harvesting, and even for frost protection. A

An outdoor sprinkler nozzle giving a fine spray.

plant that suffers no water stress will reach harvest quicker and have a greater weight than one which needs to struggle to survive. The judicious use of water will help to obtain ideal soil moisture levels when cultivation operations are to be done. For intensive crop production in most areas of the country, irrigation is a necessity as soil moisture deficiency can build up to levels which will reduce plant growth in seventeen years out of twenty.

There is a range of systems available to the grower, and the advantages and disadvantages of each need to be assessed and matched to individual requirements. The main considerations other than cost are droplet size and the area needed to be covered. Droplet size is important as it can have a marked effect on the soil. Some types of soil can be prone to capping, particularly those with poor structure or low levels of organic matter. The effect of large droplets falling on the soil will pulverise the surface, destroying the crumb structure. This can happen to the extent that it can prevent the germination of seeds. Those types of irrigation that apply water in large drops are best suited to established crops. Finer types of irrigation are best for seed-bed work and on sandy soils. Large droplet size has the advantage that it is less likely to be blown by wind, and therefore can give a more even spread over the field.

When choosing an irrigation system the area that will be irrigated and the frequency of irrigation need to be considered. The other major factor is to match pump output – both in terms of pressure and volume

A mobile irrigation pump.

of water – to the irrigation type. Rain-gun-type irrigators require a relatively high pressure (approximately 70 p.s.i.) to operate effectively, whereas oscillating spraylines can work effectively at much lower pressures.

Thought must be given to the source of water when planning an irrigation system. Mains water is expensive, and must have a ball valve or some other arrangement to prevent 'suck back' into the water supply. This is imperative if liquid feeding is to be done. Boreholes and streams may provide good sources of supply in some areas. Check with the local Water Authority. An extraction licence will be required in any case.

Having located a suitable source of water, some form of reservoir will be needed. These can be sunk into the ground and may need lining with butyl rubber to ensure they are watertight. Alternatively, storage tanks – also lined – can be erected. A pump of suitable output will be needed; these can be electric or motor driven. It is possible to get pumps to fit on to the power take-off on the tractor. This latter could cause problems when the tractor is needed for other operations at these times.

Some form of main pipe will be needed; when laying this, it is best buried below normal cultivation depth and ideally below subsoiling depth to minimise damage. Always use the largest size of main you can afford as future requirements cannot always be assessed accurately. There is nothing worse than having to replace a main with one of a larger bore two years after laying it. The pressure drop in large bore pipes is also less over a given distance. The system should be as simple as

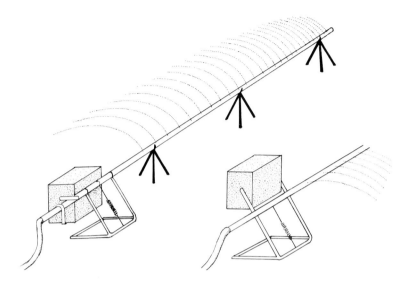

*Fig 12   Bucket oscillator and sprayline. The bucket
fills and empties which turns the spraylines.*

possible, with minimum bends as pressure is lost due to friction each
time it changes direction.

Types of Irrigation

Oscillating Spraylines

These produce a relatively small droplet and are ideal for use on seed-
beds. There are two basic types available: bucket and pump versions.

The system illustrated in Fig 12 is mobile; it does take some time to
move, but I was able to move one 150 feet (38 m) to the next site in
about twenty minutes. This system can be used on most field crops
effectively. The spread can be sensitive to the wind and frequency of
pathways may need to take this into account. If using this system,
ensure that the water supply is well filtered as weed seeds and other
debris can easily block the small jets and result in uneven application.

Portable Grid Systems

These can be purchased either in rigid plastic or aluminium pipes.
Combining this with self-sealing risers enables the spray heads to be

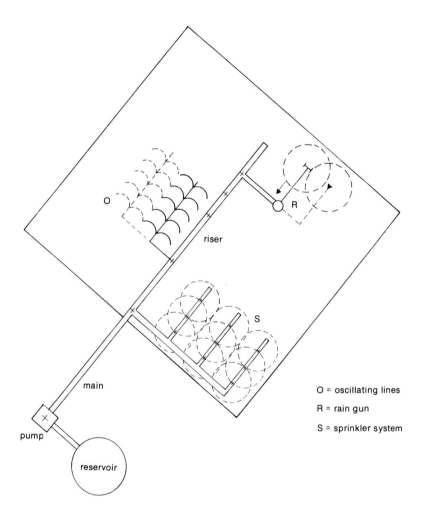

*Fig 13   Layout and comparative areas covered by
oscillating spraylines, sprinklers and rain guns.*

moved much quicker than the previous system. It is also possible to
move the sprinklers while the system is operating but you do tend to
have a shower. If finances allow, the purchase of sufficient pipes, so they
can be put down for the season and only removed for cultivations, etc.,
is to be recommended. The droplet size will depend on the nozzle type.

The most common form is the periot type or 'ticking nozzle', where the droplets will normally be medium-sized. They will not be too sensitive to wind blow but can cause some capping on certain soils.

Rain Guns

Under this heading I include the reel type of irrigators. These can be found in a range of sizes and capacities, some of which are suitable for the market garden. They can be moved relatively easily from one site to another but require a good pipe pressure. They also cover a wider area. The droplet size will be large and could cause problems but will be the least sensitive to wind.

When using irrigation it is worth checking occasionally the amount of water being applied. This can easily be done by placing a rain gauge on the area being irrigated. It is also beneficial to try to irrigate either late in the day or early in the morning. This enables the soil to absorb the maximum amount of water and reduces losses from evaporation. Watering during the heat of the day in summer can cause crop damage from scorch and it will take much longer to wet the soil effectively.

7 Protective Structures and Films

The use of protection, be it glass or polythene, can extend the range of crops grown and the harvesting season. There cannot be many holdings that do not use some form of protection, as it enables the grower to provide near-ideal conditions for the crop. It gives better returns and for certain crops allows greater control of the plant, e.g. tomatoes.

Each type of protection has its own characteristics and needs to be treated differently. Glass retains radiant heat whereas polythene does not, but poly-tunnels have a higher humidity. Modern developments in poly-tunnel design and construction now allow ridge ventilation but remove some of the main advantage of polythene, i.e. cost. All types of protective structure have a part to play on the modern market garden and will be considered in turn.

GLASSHOUSES

The modern trend in commercial glasshouse design is towards the 'Venlo' type. This is a glasshouse that can be constructed in a variety of materials but has bays 10½ feet (3.2m) wide. The eve height is normally 8½ feet (2.6m) but can be varied to suit individual needs. The number of bays can be almost infinite, and these houses can be added to at a later date if finances allow.

Most modern glasshouses are built round a steel structure and use aluminium glazing bars. The older types may have a steel frame and wooden glazing bars.

Glasshouses are expensive and can be bought either new or second-hand. There are some firms in this country which specialise in second-hand Dutch glass which is very competitively priced, and may work out cheaper than some of the more complex polythene tunnels.

When considering what type of glasshouse to buy, several factors need to be considered. Ventilation is important. Many glasshouses do not have a really adequate area of ventilation. Ideally, ventilators should be situated in the ridge of each bay, and both sides of the ridge should have openings. The total area of ventilators should approximate to one

A modern, aluminium-framed, 'Venlo'-type glass-house. A two-bay version.

third of the floor area; in practice, this area is often reduced to one sixth.

Another important consideration is light transmission. This may well relate to the use to which the glasshouse is to be put. Where crops are to be grown in winter when low light levels occur, any loss of available light will considerably reduce crop growth. For this reason, I would recommend the use of steel-framed houses with aluminium

Automation of vents in a glasshouse.

93

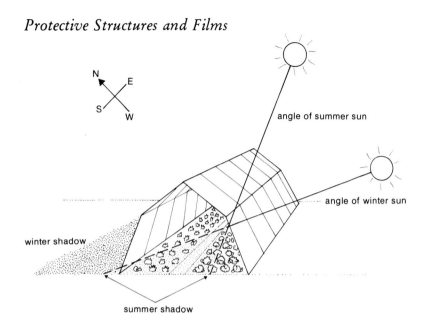

Fig 14 The shadow cast by a glasshouse in winter and summer.

glazing bars. There is less structure in these types to shade the crop. When growing winter crops in wooden houses, painting the structure white will give more reflected light. Orientation of the glasshouse can also play a major part in light transmission. When the grower's main criteria is maximum winter light, the house should be orientated in such a way that the ridge casts least shadow over the crop. This will not be possible for large multi-bay units.

The crops to be grown under glass will play a major part in deciding what type is best. A tall, trained crop (tomatoes and cucumbers) will need good height and substantial crop supports. The ends of the glasshouse need to be well braced to prevent distortion.

A heating system of some sort may be desirable. There are a range of types which can use a multitude of fuels. On small areas either gas or oil fired, free standing or suspended direct air heaters are the most common. Some form of heating will be necessary if out-of-season crops are to be grown. As a guide, a glasshouse will maintain a temperature approximately 4 °F (2.5 °C) higher than outside. This is due to glass trapping the radiant heat within the structure. When older glasshouses are being used, regular maintenance of vents and glass fittings will ensure that no heat is lost through gaps. The side walls of structures can be insulated by using 'bubble plastic', which may be left up in summer.

Bubble plastic insulation on a cut flower crop (Alstroemeria).

The main disadvantage of this type of insulation is that it will reduce the amount of light entering the glasshouse. This may be improved by insulating the north side only. Wind can also dramatically reduce the amount of heat being retained. When positioning glasshouses a protected site should be selected or some form of wind-break should be positioned near the glasshouse.

Any glasshouse or poly-tunnel will need an irrigation system. Some form of overhead system of spraylines should be installed. These may use either aluminium pipes, which are expensive, or plastic, which tend to sag in the summer heat. In all cases, the nozzles should be arranged to prevent dripping onto the crop beneath. The spraylines should *not* be

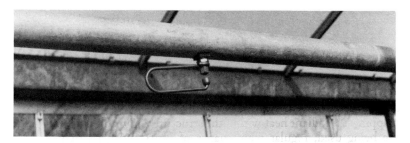

Glasshouse irrigation: aluminium spraylines and brass nozzle.

Glasshouse irrigation: plastic nozzle.

level to allow surplus water to drain down. This drain down water needs to be piped outside or wet areas will be formed in the border soil.

All valves should be readily accessible. I recommend the use of lever valves, that is those that turn off quickly, unless you like a cold shower. I also like to incorporate a hose point in each house, and the possibility of linking in a dilutor to enable liquid feeds to be applied should not be overlooked. Do not forget access. One glasshouse I inherited on my own holding only had doors at one end, and it was 120 feet (36.5m) long. Tractors and equipment can require more space to enter glasshouses than may be expected.

Roof water should be drained away, ideally to a reservoir. Some provision should certainly be made as a lot of water can be trapped on glasshouse roofs.

POLY-TUNNELS

In this section walk-in tunnels will be considered. These range in size from 14 feet (4.25m) in width upwards. They can be obtained in lengths of up to 150 feet (45.7m) and may be single span or multi-span. They are considerably cheaper than glasshouses.

The polythene skin will need replacing at regular intervals, usually every third season. When comparing various makes, the quality of the framework should be looked at in detail.

The smaller tunnels are relatively simple to erect, the ground tubes being hammered into the soil and then the steel work erected. The polythene is attached to a wooden end frame and then buried in a trench down the sides. When erecting the larger sizes more care is needed. Choose a calm day for covering. A sheet of polythene 150 × 40 feet (45.7 × 12.2m) takes some holding down. Even small covers can be impossible if there is any breeze.

As a growing environment, the poly-tunnel differs greatly from that of the glasshouse. Due to the construction, ventilation can be poor,

A typical walk-in poly-tunnel.

particularly on the larger tunnels. The use of fans to aid air flow may be necessary. Siting a tunnel on a slight slope will improve air movement, with hot air rising and cold air draining to the low end. Condensation on the cover can be problematic, though some modern films are reported to reduce this problem. When high humidity levels and a lack of ventilation are combined, real problems can occur.

The other major difference between polythene and glass is that of heat retention. Polythene, unlike glass, does not trap radiated heat. This means that at night the temperature will be very little different from that outside. The cover of a polythene tunnel will reduce the wind-chill factor, and keep the crop warmer. There are, however,

Ventilation in poly-tunnel. Note the ventilator above the door.

Spraylines in poly-tunnel. Note the hose at the end to remove the 'drain down' water.

certain circumstances when temperatures can be lower inside a tunnel. Air movement or wind, can on some nights prevent a frost occurring. The tunnel protects the crop from wind and can therefore actually encourage a frost. It may be possible to prevent this occurring by opening up the vents (doors, etc.) to allow air movement.

Erecting crop supports necessitates a careful assessment of potential loadings on the frame. It may be best to erect a separate system of them.

Poly-tunnels can be heated by similar systems to those mentioned in the section on glasshouses (*see* page 92). Double-skinned tunnels will reduce heat losses and condensation. The two layers of polythene will reduce the light levels. Irrigation will be needed, and the same systems that are available for glass are suitable under polythene. Suffice it to say, you must ensure that the water reaches all parts of the floor.

While talking of water, roof run-off must be considered. It is much more difficult to collect than that off glasshouses, except for multi-spans. If several single-span tunnels are erected close to each other drainage may be a real problem, particularly on sloping sites.

A wide range of crops can be grown successfully under polythene, within the limitations we have explored. Poly-tunnels have become popular due to the costs involved. It is possible, for the same amount of capital investment, to erect probably double the amount of protective cover by opting for polythene as opposed to glass.

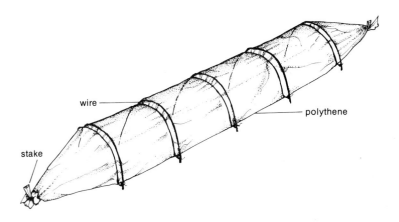

Fig 15   Low poly-tunnel or cloche.

LOW POLYTHENE TUNNELS

These may be equated more accurately to polythene cloches. These can readily be made up on the market garden. They are cheap, easy to use and will advance growth. They can be used to cover a wide range of crops usually grown outside. Low polythene tunnels have been used with success on strawberries, lettuces, French beans, early marrows and courgettes, in fact on most low-growing crops. These tunnels are usually 3 feet (91 cm) wide and consist of a series of wire loops over which the polythene is laid. The polythene sheet is secured at each end by a stake and strings hold the polythene down to the hoops.

Ventilating such tunnels is easy, as the polythene can be lifted on one or both sides and covered again at night. No special irrigation is required as water will work its way in from the sides. Due to their construction, the application of crop protection chemicals can be time consuming and weeding has to be done by hand. The covers may be able to be used several times and the hoops should last for years.

A simple method can be employed to make the hoops, as follows: For a tunnel to cover approximately 3 feet (91 cm) width, you will need:

1. A roll of polythene 4 feet (1.22 m) wide.
2. A plank approximately 6 feet (1.83 m) long.
3. Two large bolts or cotton reels.
4. A roll of stout wire.
5. Wire cutters.
6. Staples.

Low polythene tunnel with sides raised for ventilation.

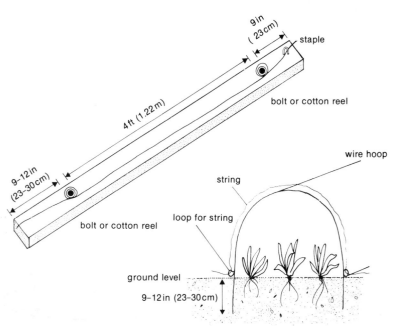

Fig 16 Board for making up hoop for low poly-tunnels, and hoop erected.

The plank should be set up as in Fig 16. The end of the wire is inserted in the staple. The wire is then wound once round the first bolt, taken straight across to the second bolt and wound round that, and finally cut off at the end of the plank. The finished hoop should look like the one illustrated in Fig 16.

When erecting the hoops, ensure the string loops are on the outside. For added security a string can be tied from one end to the other and attached to the top of each hoop. Two hoops should be used at each end to prevent a collapse.

FLOATING FILMS

This is the latest development in protective cropping. The technique and the range of films have developed over the last five or so years, and now large areas are covered using this technique. Simply, it is a large sheet of polythene or other material laid directly over the soil or young crop. Due to the protection and soil-warming properties, the crop will become established and grow faster than it would uncovered. This can improve the returns for that part of the crop, ease the spread of harvesting, and it has been claimed that some types of film exclude aphids.

Floating film, perforated polythene type.

Protective Structures and Films

Condensation on floating film.

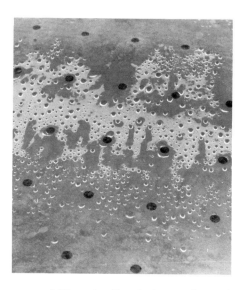

There are now two basic types of film, the first being perforated polythene sheet, the second a sort of loosely woven material not unlike nappy liners. Both types allow some air movement, which is important as it prevents the build-up of very high temperatures and reduces humidity levels. Water, from rain or irrigation, can also pass through both types of film.

The films can be laid by hand or machine and come in a range of widths, the narrower ones commonly being machine laid. Once down they are surprisingly secure as the wind tends to push them against the ground, until an edge lifts and the wind can get underneath.

If the sheets are being used to cover a crop from sowing, the drills should be sunk into the seed-bed to protect the emerging seedling from damage. As the crop grows, the sheet is lifted. The outer row will suffer more damage than the inner rows and may in fact never reach a harvestable quality. For this reason, the use of wide sheets will reduce the proportion of outside rows.

The one critical operation when using floating films is to time their removal correctly. It is likely that more damage can be done by removing the film too late rather than too early. The time of harvest can be brought forward by up to two weeks by using these floating films. The lifting operation should be done with care so none of the sheet is left in the ground. It may be possible to re-use the sheet of film if damage has been kept to a minimum.

One drawback is that it is difficult to apply crop protection chemicals

102

once the sheet is in place, so all possible spraying operations should be done before laying the film. It is not unknown for the film to be lifted to allow spraying, and then replaced.

Each of these four types of protection has its place in the modern market garden. They all work well within their own limitations and I used all four types on my own seven-acre market garden.

8 Pest, Disease and Weed Control

In this chapter the principles of pest, disease and weed control will be discussed. Some specific pests and diseases have been mentioned, with some indication of control measures, in Chapter 5 on crops. It is not possible to cover the full range of problems for such a wide range of crops as we have already looked at. It is, however, well worth considering the basic principles of pest and disease control, many of which make up what can only be described as good husbandry.

The primary aim of any grower must be to produce the crops required in such a way as to reduce to a minimum the likelihood of pest and disease infestation. This is made more difficult by the fact that crops will be grown intensively. The greater the numbers of an individual crop grown, the greater are the chances of a problem specific to that crop occurring.

For the sake of simplicity, this subject of pest and disease control will be divided into two sections, namely preventative measures and control measures.

PREVENTATIVE MEASURES

It is in this section we will consider good husbandry techniques.

First and foremost, the grower's aim must be to produce a healthy plant. To do this, the plant requires its own ideal environmental conditions. The nature of the soil in which it is growing is of vital importance, as discussed in Chapter 3. Providing the crop with a soil of good balanced nutrient status, with a good structure and no problems of impeded drainage or compaction, will enable the plant to be healthy. A healthy plant, like a healthy person, is more able to fight off pest and disease attacks from its own resources. This is a point of view that has been argued quite strongly by the modern organic movement and has tended to be overlooked by the rest of the horticultural community. The modern market gardener asks a lot of his basic resource, the soil. He or she takes a lot out and needs to put a lot in to maintain the natural balance.

One way this balance can be helped is by working a good rotation. The use of a rotation will do two things: it will help balance the demands on the soil by the crops and also help prevent the build-up of soil-borne pathogens. When considering this latter point, the longer the time interval between similar crops the better. The one real problem with rotations is that for reasons of economics and marketing opportunity the grower may have decided on a cropping schedule which involves specialisation in certain crops. I would recommend that, if at all possible, some form of rotation be practised. It may be that certain areas can be left fallow or a green manure crop can be grown on areas not actively being cropped. When this is not possible the grower should, at the least, be aware of the possible pitfalls.

After rotation and the soil, the general growing conditions need to be considered. It is not possible to change the other environmental conditions under which the crop is grown except by the use of protective structures (as discussed in Chapter 7). To help reduce the incidence of pathogen infection in these areas, the grower should ensure the conditions are such that they do not encourage the build-up of the pathogens. This can be done by ensuring that there is adequate ventilation and air movement through the crop. Such techniques as watering early in the day so the crop canopy (leaves) dries out before nightfall will help.

When carrying out any cultural operations, such as weed control or training, every effort should be made to keep any damage to the crop to a minimum. This may seem commonsense, but while a broken leaf may not reduce the quality of the crop, it offers an easy entry point for disease.

A crop that has been grown 'soft' is more tempting to sap-sucking insects such as aphids, and infestations may be reduced by growing the crop with a 'harder' regime. Whilst on the subject of aphids, there has in the past been a move to eradicate the pests' winter host. Latest research shows that removal of the spindle-tree (*Euonymus europeus*) from the hedgerows does not reduce the incidence of black bean aphid in broad beans. Keeping a varied flora and fauna on the uncropped areas of the holding will encourage the presence of natural predators in the area and so help the grower's efforts to control pests. Encouraging ladybirds is the classic example.

Next on the list of preventative measures must be hygiene. This has been mentioned many times earlier in this book. So what do we mean by hygiene? Put simply, it is the elimination of any potential source of infection. The most obvious source is crop debris left after harvest. This should not be allowed to remain for any length of time. Depending on what problems the previous crop has encountered, it may be sufficient to incorporate it quickly into the soil. If it is to be removed from the growing area it should not be dumped close to other similar crops. All rubbish tips should be regularly maintained to prevent fungal spores being carried onto healthy crops. This can be done by composting the organic matter, covering with soil, or in some cases burning may be necessary.

Water sources should also be protected from infection where possible. That ideal site for a rubbish dump by the reservoir is not so ideal after all.

All machinery and equipment should be cleaned regularly. This is particularly important if it has been working on or in an infected soil. In fact, soil-borne diseases such as club root have been known to move from one holding to another on tractor wheels and even on boots of visitors. I don't mean to be alarmist or advocate a policy of isolation but prevention is cheaper than cure.

All propagating materials should be regularly sterilised as should the structure of any glasshouse or poly-tunnel. In intensively cropped areas it can be beneficial to sterilise the soil as a routine preventative measure. Chemicals can be used for this, or steam sterilisation can be used. Steam is becoming a possibility again, since the production of small, portable steam generators.

Finally, with regard to hygiene, do not overlook the work force and their tools, such as knives. Several tomato diseases can be spread by infected knives or fingers.

The use of clean seed and plant material can go a long way in preventing the introduction of any pests and diseases. All seed suppliers offer a range of treatments on their seeds. Whilst increasing the cost of the seed initially, it may well be the cheapest measure of controlling certain diseases. This avenue is not open to the organic grower and most seed houses now supply a range of varieties available untreated. If in doubt about seed treatments, a chat with the local seed salesman will be worthwhile. In many respects the same is true of plant suppliers. Some spray their plants as a matter of course; others do not. If you have specific requirements, chat to them and find out. While it is in the interest of the plant supplier to send out only clean plants, this does not always happen, and any plants bought in should be inspected closely at the time of delivery. When choosing what variety of plant to grow it can be beneficial to choose one which is resistant to certain pests and diseases. I have already mentioned the parsnip Avon Resistor (*see* Chapter 5, page 61). Most modern tomato varieties are TMV resistant or greenback resistant. Many varieties of lettuce are resistant to certain strains of mildew. By careful selection of resistant varieties many problems can be eliminated. Plant breeders are working all the time to improve this natural form of prevention so keep up-to-date and some problems may be prevented.

Finally, as far as preventative measures go, it may be necessary to take some quite drastic action to prevent the occurrence of outbreaks of certain pests and diseases. This may mean not growing the crop at all, or possibly growing in an isolated medium such as peat modules or grow bags. This is normally only practised where a specific soil problem has become uncontrollable or if the grower needs to be able to control the plant more precisely.

CONTROL MEASURES

It is likely that however careful a grower is with his preventative measures he will suffer some, if not many, pathogen infestations. When this occurs some form of curative measure needs to be taken. There are three basic ways the problems can be attacked. These are by using:

1. Chemical pesticides.
2. Biological means.
3. An integrated system combining both 1 and 2 above.

Before we look at each of these in turn, the grower needs to know when to instigate a method of control. In most cases the problem can be controlled more easily if the symptoms are seen soon enough. To be sure of doing this, all crops should be regularly inspected. In fact, this can be a very relaxing and worthwhile job. Walking through the crops as regularly as possible, inspecting for pathogen attack, and noting crop growth and development enable the grower to keep on top of problems. This is preferable to waiting for a problem to develop and then rushing about trying to solve it. Staff should be instructed to look for anything unusual, even to the extent of offering a reward for the first whitefly!

Chemical Pesticides

Control of pathogen attack by means of chemicals is by far the most common method. It is not intended in this section to recommend specific chemicals for problems; for this, contact the local ADAS office or consult the relevant publications, or talk to the local supplier who can often be very helpful.

Before going any further, we need to be aware of the recent changes in the law concerning the application of pesticides. This has already been mentioned in the section on sprayers (*see* Chapter 6, page 82). The legal requirements I noted there, albeit very briefly, do have far-reaching implications for the grower. There has been a number of Acts of Parliament which regulate the use of pesticides, the latest ones (at the time of writing) being the Food and Environment Protection Act 1985 (FEPA), and the Control of Pesticide Regulations 1986. These Acts and Regulations affect all who supply and use pesticides. For the grower, these regulations basically mean that:

1.   Chemicals can only be used in *approved* circumstances; the details should be on the product label.
2.   Anyone applying such chemicals should have a recognised Certificate of Competence issued by the National Proficiency Test Council for both the foundation module and for the specific method of application being used. There are some exceptions to this.

For further information and for advice, contact your local Agricultural Training Board or agricultural college. These regulations are designed to protect the grower, his or her staff, the public and the environment.
    Anyone completing the necessary course and successfully gaining a certificate should be a competent operator. I would, however, like to make some points about chemical applications.

Pathways should allow access to all parts of the crop when spraying (lettuce seedlings).

When deciding how best to apply chemicals, thought should be given to how to get the chemical where it will be effective. Many pests live on the undersides of leaves and spray droplets need to be directed here.

Fine sprays will enable maximum amounts of chemical to be absorbed by the plant. Therefore, droplet size will play an important part in successful application. All equipment used should be well maintained and calibrated (*see* Chapter 6, page 82).

When using chemicals to control both pests and diseases, a range of different chemicals should be used. This prevents the build-up of resistant strains of pathogens. Strains of lettuce mildew have developed resistance to Metalaxyl, and some pests are resistant to certain insecticides. By ringing the changes this can be prevented. The chemical should be changed each time the crop is sprayed and it is usual practice to have a minimum of three chemicals in the spray programme. This holds good for both insecticides and fungicides.

Pathogen control does not stop with harvest. It is sensible to fumigate all crop debris before removal from protective structures as moving crop rubbish will disperse fungal spores and insects. Do not forget rubbish tips as well. Regular fumigation or sterilisation of the physical structure of glasshouses and tunnels should be carried out to prevent pests and diseases overwintering and infecting future crops. I used commercial formaldehyde and appropriate protective clothing for this. The soil will need attention too.

When using chemicals, it goes without saying that all harvest intervals must be strictly adhered to. The time of application is also important. All spraying operations should be carried out in cool weather. This helps to prevent any damage to the crop from scorch. The best time is in the evening, when insects are often still active and the temperature has dropped. It allows the plant to absorb the maximum amount of chemical and none is lost by volatilisation.

Chemical control can be very effective, and provides the grower with a wide range of weapons to protect his crops and ensure an economic return for his labours.

Biological Control

At present, the range of pests that can be controlled biologically is limited. Research is continuing to find new methods. The pests for which biological control is currently available include red spider mites, whiteflies, leaf miners, mealy bugs, caterpillars and some aphids.

How does biological control work? The pest is controlled by introducing a predator or parasite which is specific to that pest. This predator or parasite then kills off the pest. In practice, this means that the pest can be kept to acceptable levels. It is possible to eradicate the pest but, as we shall see, this may not be desirable. Before progressing any further, it should be said that biological control works best under glass or polythene. The predators tend to prefer higher temperatures than can normally be expected outdoors in the UK, but they can still be effective there.

Biological control can be used in two ways. The aim of the first system is to prevent a build-up of the pest over the growing season of the crop. This is done by introducing the pest and predator at an early stage in the crop growth. The number of predators will control the pest to acceptable levels. If a sudden infestation occurs from outside sources, the predator is ready to multiply to deal with it. In severe attacks, a further introduction of predator may be necessary to achieve control quickly. When using this system, regular monitoring of both pest and predator is needed to ensure that the predator is always present.

The second system is a relatively short-term control. Introduction of the predator is left until the pest has infected the crop. The rate of introduction of the predator is related to the level of infestation, and rapid control is possible. I have used this method with success when an infestation of red spider mite on a cold tomato crop took place in August.

These control methods apply where insect predator or parasite is used to control the pest, i.e. red spider mite, whitefly and leaf miner. Biological control of caterpillars is included in the FEPA regulations. It can be applied either as routine sprays or in response to damage. The aphid control *Verticillium lecanii* is a fungus and is applied the same way. When applying sprays of these two controls, high-volume sprays should be used to ensure a thorough wetting of the plants.

Other means of biological control can be used. Coloured boards smeared with a sticky substance such as petroleum jelly can be suspended above the crop. Bright yellow is the colour most insects find attractive; they fly to the board and become stuck to it. These boards are good indicators of a pest infection and can be used as an early warning system.

Many pests are attracted to the plant by smell, for example carrot- and cabbage-root fly. It may be possible to mask the smell of the crop, and research is under way to establish if it is possible to use this mechanism for pest control. Companion planting often works in this way. The planting of marigolds near crops affected by aphid and whitefly can be useful. The smell is strong and masks the scent of the normal host plant, so the aphid and whitefly are not attracted to it. We get put off by strong smells, and insects are more sensitive to odour than we are!

Tagetes being used to discourage insects from a brassica crop.

As we have seen, biological means for control of pathogens is effective but limited. It has advantages over chemical control in that there is no harvest interval, no damage to crops and less time – if any – is spent spraying. Also, it cannot pollute or damage the environment or be harmful to staff or customers.

Integrated Control

This system utilises both biological and chemical control and so combines these two methods. Its use is becoming more widespread as the common pests can be controlled biologically, but chemicals are used where no biological control is possible. In this system, careful selection of the chemicals is vital as many will adversely affect the predator. The best advice on what chemicals can be used safely can be obtained from the suppliers of biological control agents (*see* Appendix 2).

Weed Control

As with pest control, weeds can be controlled both by chemical and natural means. Well, by natural perhaps I should say mechanical. Weed control equipment was covered in Chapter 6, page 80. Suffice it to say here that there is a wide range of mechanical weeders available.

When considering chemical weed control we need to ascertain the mode of action of these herbicides.

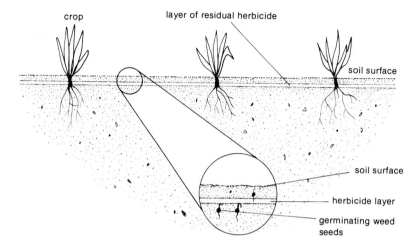

Fig 17 Layer of residual herbicide in damp soil.

Contact Herbicides   These, as the name implies, kill on contact. The exact mode of action varies slightly but basically the chemical will kill that part it touches. Paraquat is a good example. It will not kill the roots of plants such as dock or thistle and is also totally non-selective in action. By that I mean it will kill weed and crop, grass and broad-leaf alike.

Translocated Herbicides   In this group, the active ingredient is absorbed by the plant and translocated to all parts. This has the great advantage that the roots will be killed as well as the aerial parts. Round-up is an example of this kind of herbicide.

Residual Herbicides   These are so called because the residues stay in the soil, and they can work in a variety of ways. They may prevent germination of weed seedlings. The chemical can be absorbed by the roots of young weeds and hence kill them. Some residual herbicides also have contact action. With these residual herbicides not all weeds may be controlled by the chemical. Care is needed in selecting the right herbicide. When doing so the following needs to be taken into account:

1.   The weeds species that are the problem (weed spectrum).
2.   The crop plant; it may also be susceptible.
3.   The soil type and moisture content.

This latter point is important when it comes to applying the chemical.

112

Polythene mulch used for weed control in bed system for strawberry growing.

Pest, Disease and Weed Control

Soil-acting herbicides are spread in moist soil and form a layer just below the surface. This will not occur if the soil surface is too rough or too dry. Once a soil-acting herbicide is applied, the ground should not be disturbed. It is possible to apply such chemicals after planting, but when doing so great care should be taken to ensure the correct dose is applied or crop damage will result.

Selective Herbicides   These may be one of the above types. They are called selective because they only kill certain kinds of plants and will leave the crop undamaged. Hormone herbicides can be classified as selective as they do not work on grasses and cereals but kill broad-leafed plants. Note that tomatoes are very sensitive to hormone herbicides and damage can easily occur from spray drift.

Total Herbicides   These kill all plants and should only be used in extreme cases as the chemical can remain in the soil for several seasons.

All chemical herbicides are covered by the FEPA regulations. I have heard of only one biological weed control – not sheep or goats but *Tagetes minutiae* which is reputed to kill ground elder. I would be interested to know if it works.

There are other forms of weed control. Mulches can be used; these may consist of polythene, special paper, straw, etc. The idea of mulching is to cover the soil surface with a layer of a sterile medium. It can act as a physical barrier, such as black polythene, or be just too deep for weeds to push their way through, such as straw. Mulches also have the advantage of retaining moisture and can help improve soil temperatures. Weeds can also be suppressed by the use of a low-growing green manure crop such as clover. This will protect the soil from damage, can be incorporated after cropping and can generally help fertility.

Effective pest, disease and weed control will be a combination of many factors. Good general husbandry, a fertile soil, good hygiene and careful selection of varieties will help to prevent infestations. When these do occur, a range of options is open to the grower. Under protected structures the use of biological means, either on their own, or integrated with chemicals are very effective and have advantages. Chemicals themselves can be extremely effective. No one system is foolproof, and the grower must adopt that which is best suited to his or her own methods and preferences. There are a great many potential problems from pests, diseases and weeds, but thankfully usually only a handful occur each season. It is not often a crop is a total loss thanks to the modern methods of control available today.

9 Management

To run a successful enterprise the owner or manager, as I shall call him or her, has to have a wide range of abilities. We all have a flair for particular things, and any skills that are needed can be bought in. In many respects, financial advice is one of the easiest to buy. A desire to succeed, combined with enthusiasm and an ability to learn will go a long way. The grower should pay attention to detail in all aspects of the business. Most people who like getting their hands dirty do not like the paperwork but it needs to be done; keeping up to date so you can see where you are financially is just as important as being on top of the growing side. All this may sound a bit daunting but don't let it put you off. It certainly will not be easy, it will involve long hours and the financial rewards may not always be good, but there is a satisfaction in producing good crops and seeing the results of your labours that is like no other.

RESOURCES

Making the most of all available resources really sums up a manager's ability. In Chapter 3 we looked at managing that prime resource, the soil. Marketing possibilities have also been discussed. In the chapter on mechanisation, financial implications were raised. All these, plus the manager's own ability, and that of the staff, must be added together to produce an efficient, well-run, profitable organisation. To make the most of that unique blend, each of the resources must be appraised. When the marketing opportunities, soil type and site are considered any decision regarding cropping may be made.

The manager's own abilities and those of the staff can be improved by training in the areas of weakness, so there should not really be a problem here. The other major hurdle to overcome is finance. To make the most of what finance is available plans should be drawn up.

PLANNING

It is surprising how many successful holdings have just grown haphazardly. Life can be made much easier with careful planning. By this I mean all types of planning.

A cropping plan needs to be made out for each season. Records should be kept so it can be modified in future to gain maximum advantage. Cropping plans very rarely work out in practice due to the vagaries of our weather. If you do have an organised plan, the effects of a delayed crop can be seen and assessed early. It may indicate that changes should be made later in the season, and planning for these changes now can turn them to advantage.

A development plan for the holding is also extremely useful and should cover a number of years: the following year in detail, provisional plans for the next four or five years and a long-term plan for ten or more years hence. This prevents expensive mistakes, such as installing structures or irrigation mains where they will not be needed in future.

These long-term plans of physical resources should be linked in with financial plans. If you are borrowing money, your bank will need to know your plans anyway, so projecting into the future should not be difficult.

While on the subject of planning, we should consider local authority planning regulations. Planning by-laws vary round the country, and permission may or may not be required for the erection of glasshouses, packing sheds, poly-tunnels and the like. If you are considering opening a farm shop it is often the case that planning consent is not required if only your own produce is sold. Rates may be affected as well if a retail outlet is envisaged. If you are in any doubt about local regulations contact the local authority concerned.

When considering retail outlets on the holding, access to and from the road must be considered. What might be acceptable for your own drive may not be acceptable for access by the general public. The Highways department of your County or Regional Council can help in this.

When a plan is produced it enables the manager to develop a basic strategy to achieve the goals by utilising the resources available. It can be modified in the light of experience and varying circumstances; the goals may change but with careful planning should still be within reach.

Access and parking at a farm shop.

FINANCE

In the previous section financial planning was mentioned, but it is also necessary to keep financial records. For one thing, the Inland Revenue and VAT require certain records and all financial details to be kept. The manager, to be effective, needs to know exactly where his business stands at any one time. The annual balance sheets, prepared some months after the end of the financial year, are not good enough to base sound financial judgement on. The manager will be making decisions all the time, and the more information he has at his fingertips the more logical will be his decisions.

It is possible to formulate a relatively simple system of bookkeeping which will keep both the Inland Revenue and VAT happy and still provide the financial information the manager requires. In fact, modern desk-top computers can be an asset; software packages are available for cropping details, etc. If you are happy using a computer, it has advantages. If you cannot stand these modern gadgets, stick to a system you are happy with. Whatever method is adopted, it should be easy and convenient. If this is so, it is much more likely to be kept up to date.

While on the subject of finance, we cannot ignore the fact that most businesses use borrowed money at some stage. It may be for capital

117

investment or for working capital to tide one over the quiet period. One word of caution: do not over-borrow. The returns from growing are not the greatest, and interest rates vary. The ability to repay and live needs to be looked at very carefully. Banks are the most usual source of finance but not the only ones. This whole subject of finance and financial management is a very complex one. There are very good books and courses available to help the grower gain a greater understanding and develop skills in this area.

INSURANCE

Insurance is available to cover a wide range of risks. Public liability and employer's liability are both likely to be necessary. Some structures may be insurable against a range of risks. Polythene tunnels are one area which can be difficult. It is normally possible to insure the framework but not the cover. Glasshouses can be insured and when insuring them, be sure to include cover to pay for removal of glass from the site. In the event of major damage, it can take many man-hours to remove all the broken glass. Crops can be covered for certain risks. It may be possible to insure glasshouse crops against heating system breakdown, etc. but it will be expensive. What may be cheaper is to insure against potential loss of income, though outdoor crops are not usually insurable. It goes without saying that all vehicles and equipment should be covered.

TRAINING

This, perhaps, seems an unusual section to include in a chapter on management, and I have already mentioned it several times, but great improvements can be made in the business by identifying where there are skill shortages and rectifying these.

At the start of this chapter I suggested that a grower should have a wide range of skills. Very few people will be good at all aspects of running a market garden. Once the areas of weakness have been identified try to find a training course to help develop skills in these particular areas. There are two main sources of training throughout the country. These are the Agricultural Training Board or ATB and the agricultural and horticultural colleges. There are grower groups and associations that may run their own training courses as well. Most of these can be found in the Yellow Pages under Training. They all run a vast range of courses ranging from basic horticultural skills up to longer management courses. The ATB local training officer may be able to help develop a complete training programme for the individual business. Most courses are organised during the period from October to March, when life tends to be a little quieter.

Running a successful business depends on many things, as we have seen. It is not possible in a book such as this to cover all aspects in detail. The owner/manager will have a wide range of skills already and it is relatively easy to gain new ones by attending any of the excellent training courses on offer today. Careful planning and attention to detail combined with optimism and enthusiasm will get you by. There is nothing quite like growing – it is very satisfying. Good luck.

Appendix 1

LIMING MATERIALS

Calcium carbonate   Soft chalk, limestone.
Most common liming material, cheap and safe, readily available.

Calcium Oxide   Quicklime, burnt lime, cob lime or caustic lime.
Not used much. It burns and is a fire risk.

Calcium Hydroxide   Hydrated lime or slaked lime.
Fine white powder. Commonly used in composts. High neutralising value. Works relatively quickly.

Magnesian Lime   Dolomitic limestone.
Used in composts and under glass as it includes magnesium, but it acts slowly.

Appendix 2

SUPPLIERS OF BIOLOGICAL CONTROL AGENTS

Brinkmans
Spur Road
Quarry Lane
Chichester
West Sussex PO19 2RP
Tel. 0243 531666

Koppert UK Ltd
PO Box 43
Tunbridge Wells
Kent TN2 5BY
Tel. 0892 36807

Natural Pest Control
Yapton Road
Barnham
Bognor Regis
West Sussex PO22 0BX
Tel. 0243 553250

Bibliography

ADAS *Plant Physiological Disorders* reference book no. 223 (HMSO, 1985)

ADAS *Plastic Film Covers for Vegetable Production* booklet no. 2434 (HMSO, 1984)

ADAS *Propagating and Transplanting Vegetables* reference book no. 344 (HMSO, 1981)

Adams, C.R., Bamford, D.D., and Early, M.P. *Principles of Horticulture* (Heinemann, 1984)

Arden-Clarke, C., and Hodges, P. *Soil Erosion: the Answer lies in Organic Farming* (*New Scientist*, 12 February 1987)

Bould, C., Hewitt, E.J., and Needham, P. *Diagnosis of Mineral Disorders in Plants* vol. 1, *Principles* (HMSO, 1983)

Davies, B., Eagle, D., and Finney, B. *Soil Management* (Farming Press, second impression 1979)

Fletcher, J.T. *Diseases of Greenhouse Plants* (Longman, 1984)

Grower Guides, no. 3: *Peppers and Aubergines*; no. 4: *Farm Sales and Pick-Your-Own; no. 5: Raspberries*; no. 6: *Celery*; no. 7: *Plastic Mulches for Vegetable Production*; no. 12: *Money in Growing*

Maddox, H. *Your Garden Soil* (David & Charles, 1974)

Robertson, J. *Mechanising Vegetable Production* (Farming Press, 1974)

Sarjent, M.J. *Economics in Horticulture* (The Macmillan Press Ltd., 1973)

Scaife, A., and Turner, M. *Diagnosis of Mineral Disorders in Plants* vol. 2, *Vegetables* (HMSO, 1983)

Schering *Working with Pesticides Guide: the Regulations and your Responsibilities* (ATB)

Webber, R. *Market Gardening* (David & Charles, 1972)

Index

Italic numerals denote page numbers of illustrations.

125

Index